THE CONTINUUM

A Critical Examination of the Foundation of Analysis

Hermann Weyl

Translated· by

Stephen Pollard & Thomas Bole

DOVER PUBLICATIONS, INC.
New York

Bibliographical Note

This Dover edition, first published in 1994, is an unabridged and corrected republication of the work first published The Thomas Jefferson University Press, Kirksville, Missouri, in 1987.

Library of Congress Cataloging-in-Publication Data

Weyl, Hermann, 1885–1955.
[Kontinuum. English]
The continuum : a critical examination of the foundation of analysis / Hermann Weyl ; translated by Stephen Pollard & Thomas Bole.
p.   cm.
Originally published: Kirksville, MO : Thomas Jefferson University Press, c1987.
Includes bibliographical references and index.
ISBN 0-486-67982-9
1. Mathematical analysis—Foundations.   I. Title.
QA299.8.W4813   1994
515—dc20   93-50762
                                                                           CIP

Manufactured in the United States of America
Dover Publications, Inc., 31 East 2nd Street, Mineola, N.Y. 11501

Acknowledgments

Two parts of this book were originally published elsewhere. They are: an article by Hermann Weyl, "Der circulus vitiosus in der heutigen Begründung der Analysis" published in *Jahresbericht der deutschen Mathematiker-Vereinigung,* Band 28, 1919, pages 85-92; and excerpts from John Archibald Wheeler's "Hermann Weyl and the Unity of Knowledge" published in *American Scientist*, Volume 74, August 1986, pages 368 ff. The translators express their thanks and gratitude to the copyright holders for their permission to use the above publications.

Note to the Dover Edition

R. V. Schnucker and Paula Presley of The Thomas Jefferson University Press deserve special thanks. Without their energy and generosity this Dover edition would not have been possible. The translators also benefited from the advice and encouragement of many outstanding scholars, especially John Corcoran and Ignacio Angelelli. Their contributions are deeply appreciated.

Table of Contents

Foreword   ix

Introduction   xv

Preface   1

Chapter 1: Set and Function   5
Logical Section
  §1. Property, Relation, Existence   5
  §2. The Principles of the Combination of Judgments   9
  §3. Logical Inference. Axiomatic Method   14
Mathematical Section
  §4. Sets   20
  §5. The Natural Numbers: Richard's Antinomy   24
  §6. Iteration of the Mathematical Process
    The *circulus vitiosus* of Analysis   28
  §7. Principles of Substitution & Iteration   35
  §8. Definitive Formulation of the Foundations.
    Introduction of Ideal Elements   40

Chapter 2: The Concept of Number & The Continuum   51
  §1. Natural Numbers and Cardinalities   51
  §2. Fractions and Rational Numbers   58
  §3. Real Numbers   66
  §4. Sequences, Convergence Principle   74
  §5. Continuous Functions   80
  §6. The Intuitive and The Mathematical Continuum   87
  §7. Magnitudes and Their Measures   97
  §8. Curves and Surfaces   101

Appendix   109

Notes   118

Bibliography   124

Index   129

Foreword

Hermann Weyl was—is—for many of us, and for me, a friend, a teacher, and a hero. A North German who became an enthusiastic American, he was a mathematical master figure to mathematicians, and to physicists a pioneer in quantum theory and relativity and discoverer of gauge theory. He lives for us today, and will live in time to come, in his great findings, his papers and books, and his human influence.

I last knew Weyl after I last knew him. Day after day in Zurich in late 1955 he had been answering letters of congratulation and good wishes received on his seventieth birthday, walking to the mailbox, posting them, and returning home. December eighth, thus making his way homeward, he collapsed on the sidewalk and, murmuring, "Ellen," died. News of his unexpected death reached Princeton by the morning *New York Times*. Some days later our postman brought my wife and me Weyl's warm note of thanks. I like to think he sent it in that last mailing.

I first knew Weyl before I first knew him. Picture a youth of nineteen seated in a Vermont hillside pasture, at his family's summer place, with grazing cows around, studying Weyl's great book, *Theory of Groups and Quantum Mechanics*, sentence by sentence, in the original German edition, day after day, week after week. That was one student's introduction to quantum theory. And what an introduction it was! His style is that of a smiling figure on horseback, cutting a clean way through, on a beautiful path, with a swift bright sword.

Some years ago I was asked, like others, I am sure, to present to the Library of the American Philosophical Society the four books that had most influenced me. *Theory of Groups and Quantum Mechanics* was not last on my list. That book has, each time I read it, some great new message.

If I had to come up with a single word to characterize Hermann Weyl, the man, as I saw and knew him then and in the years to come, it would be that old-fashioned word, so rarely heard in our day, "nobility." I use it here not only in the dictionary sense of "showing qualities of high moral character, as courage, generosity, honor," but also in the sense of showing exceptional vision. Weyl's eloquence in pointing out the peaks of the past in the world of learning and his aptitude in discerning new peaks in newly developing fields of thought surely were part and parcel of his lifelong passion for everything that is high in nature and man.

Erect, bright-eyed, smiling Hermann Weyl I first saw in the flesh when 1937 brought me to Princeton. There I attended his lectures on the Élie Cartan calculus of differential forms and their application to electromagnetism—eloquent, simple, full of insights. Little did I dream that in thirty-five years I would be writing, in collaboration with Charles Misner and Kip Thorne, a book on gravitation, in which two chapters would be devoted to exactly that topic. At another time Weyl arranged to give a course at Princeton University on the history of mathematics. He explained to me one day that it was for him an absolute necessity to review, by lecturing, his subject of concern in all its length and breadth. Only so, he remarked, could he see the great lacunae, the places where deeper understanding is needed, where work should focus.

The man who ranged so far in his thought had mathematics as the firm backbone of his intellectual life. Distinguished as a physicist, as a philosopher, as a thinker, he was above all a great mathematician, serving as professor of mathematics from 1913 to 1930 at Zurich, from 1930 to 1933 at Göttingen, and at the

Princeton Institute for Advanced Study from October 1933 to his retirement. What thinkers and currents of thought guided Weyl into his lifework: mathematics, philosophy, physics?

"As a schoolboy," he recalls, "I came to know Kant's doctrine of the ideal character of space and time, which at once moved me powerfully." He was still torturing himself, he tells us, with Kant's *Schematismus der reinen Verstandesbegriffe* when he arrived as a university student at Göttingen. That was one year before special relativity burst on the world. What a time to arrive, just after David Hilbert, world leader of mathematics, had published his *Grundlagen der Geometrie*, breaking with Kant's predisposition for Euclidean geometry and taking up, in the great tradition of Karl Friedrich Gauss and Bernhard Riemann, the construction and properties of non-Euclidean geometries, and—having just published an important book on number theory, *Zahlbericht*—was giving absorbing lectures on *that* field of research. Philosophy! Mathematics! Physics! Each was sounding its stirring trumpet blast to an impressionable young man. Mathematics, being represented in Göttingen by its number one man, won the number one place in Weyl's heart.

Weyl tells us the impression made upon him by Hilbert's irresistible optimism, "his spiritual passion, his unshakable faith in the supreme value of science, and his firm confidence in the power of reason to find simple and clear answers to simple and clear questions." No one who in his twenties had the privilege to listen to Weyl's lectures can fail to turn around and apply to Weyl himself those very words. Neither can anyone who reads Weyl, and admires his style, fail to be reminded of Weyl's own writing by what he says of the lucidity of Hilbert: "It is as if you are on a swift walk through a sunny open landscape; you look freely around, demarcation lines and connecting roads are pointed out to you before you must brace yourself to climb the hill; then the path goes straight up, no ambling around, no detours."

Electrified by Leibniz and Kant, and under the magnetic influence of Hilbert, Weyl leaped wholeheartedly, as he later put it, into "the deep river of mathematics." That leap marked the starting point of his lifelong contributions to ever widening spheres of thought.

For the advancing army of physics, battling for many a decade with heat and sound, fields and particles, gravitation and spacetime geometry, the cavalry of mathematics, galloping out ahead, provided what it thought to be the rationale for the real number system. Encounter with the quantum has taught us, however, that we acquire our knowledge in bits; that the continuum is forever beyond our reach. Yet for daily work the concept of the continuum has been and will continue to be as indispensable for physics as it is for mathematics. In either field of endeavor, in any given enterprise, we can adopt the continuum and give up absolute logical rigor, or adopt rigor and give up the continuum, but we can't pursue both approaches at the same time in the same application.

Adopt rigor or adopt the continuum? These ways of speaking should not be counted as contradictory, but as complementary. This complementarity between the continuum and logical rigor we accept and operate with today in the realm of mathematics. The hard-won power thus to assess correctly the continuum of the natural numbers grew out of titanic struggles in the realm of mathematical logic in which Hermann Weyl took a leading part. His guidance, his insights and his wisdom shine out afresh to the English-speaking world with the publication of the present volume. The level of synthesis achieved by now in mathematics is still far beyond our reach today in physics. Happily the courageous outpost-cavalry of mathematical logic prepares the way, not only for the main cavalry that is mathematics, but also for the army that is physics, and nowhere more critically so than in its assault on the problem of existence.

Hermann Weyl has not died. His great works speak prophecy to us in this century and will continue to speak wisdom in the coming century. If we seek a single word to stand for the life and work of Hermann Weyl, what better word can we find than passion? Passion to understand the secret of existence was his, passion for that clear, luminous beauty of conception which we associate with the Greeks, passionate attachment to the community of learning, and passionate belief in the unity of knowledge.

John Archibald Wheeler
University of Texas at Austin

Introduction

Anyone who hikes in the mountains with my youngest nephew soon learns to expect sudden stops, starts, and ventures off the main path. So perhaps my experience as an uncle has left me better prepared than others to follow Hermann Weyl's intellectual excursions. Weyl had a fertile and lively mind which delighted in tangential adventures and abrupt transitions. When he applies his restless style to a topic which is quite complex enough just in itself, the effect can be dizzying—all the more so when the reader is simultaneously faced with an unusual terminology and symbolism. In the hope of making the reader's tour of *The Continuum* somewhat smoother, let me take time now to sketch (without detours) the major developments of Chapter 1—and let me do so in the current language of logic and foundational studies. This translation into the current idiom may produce occasional distortions. But there are situations where an intelligible distortion is a better guide than a more faithful but less penetrable account. And I hope the reader will grant that we face just such situations in what follows.

Let me give only the briefest attention to Section 1 of Chapter 1. Weyl launches a two-pronged attack on the corruption he detects in the foundations of analysis. In order to avoid the antinomies of set theory, he subjects the epsilon relation to type restrictions. In order to avoid the *circulus vitiosus* of impredicative definitions, he restricts the use of quantifiers in the set-theoretic comprehension scheme. In Section 1, Weyl tries to reconcile us to the general idea that fetters such as these are necessary if we are to avoid meaninglessness and contradiction. To this end he appeals to our ordinary use and understanding of language and he discusses one of the semantical paradoxes

(Grelling's). I leave it to the reader to analyze Weyl's argument in detail.

It is in Section 2 that Weyl begins to lay out his (more or less) formal system. Here Weyl introduces a family of languages whose most noteworthy feature is the absence of variables in closed formulas—more precisely, languages of the following form. (For simplicity's sake, I will, for the moment, omit the complications which attend type restrictions.)

LEXICON

Primitive Relation Symbols: 'R_1', 'R_2', . . .
Derived Relation Symbols: 'F_1', 'F_2' . . .
Individual Constants: 'c_1', 'c_2', . . .
Logical Symbols: ')', '(', '+', '−', '∗'

FORMATION RULES

1. If p is a relation symbol and $a_1, ..., a_n$ is a finite sequence each of whose terms is either an individual constant or '∗', then $pa_1, ... , a_n$ and $\bar{p}a_1, ..., a_n$ are formulas.

2. If p and s are formulas, then so are $\bar{p}$ and $(p + s)$.

In place of a full-blown truth definition for this language, let me translate some sample formulas into more famliar notation:

$$\text{`}(R_1c_1c_2 + R_2c_3)\text{'} \quad \text{means} \quad (R_1c_1c_2 \text{ v } R_2c_3)$$

$$\text{`}\overline{(R_1c_1c_2 + R_2c_3)}\text{'} \quad \text{means} \quad -(R_1c_1c_2 \text{ v } R_2c_3)$$

$$\text{`}\overline{R_1}c_1c_2\text{'} \quad \text{means} \quad -R_1c_1c_2$$

$$\text{`}R_1{**}\text{'} \quad \text{means} \quad (\exists x)(\exists y)R_1xy$$

$$\text{`}\overline{R_1}{**}\text{'} \quad \text{means} \quad (\exists x)(\exists y)-R_1xy$$

$$\text{`}\overline{R_1{**}}\text{'} \quad \text{means} \quad -(\exists x)(\exists y)R_1xy$$

Since each '∗' represents a distinct existential quantifier, we are left with no way of expressing, e.g., that $(\exists x)R_1xx$; that is, we are left without the syntactic resources for representing pronominal cross-reference—or, at least, so it seems. Weyl responds to this problem by adopting Carnapian meaning postulates (pre-Carnap, of course) which link the primitive to the derived relation symbols. For example:

> If b is an individual constant, then R_1bb is equivalent to F_1b.

We can reflect this semantic postulate at the deductive level by adopting the corresponding synonomy rule for 'R_1':

> If b is an individual constant, then from R_1bb infer F_1b —and vice-versa.

Now we can use 'F_1*' to express that $(\exists x)\, R_1xx$. And we can reproduce the inference from, say, '$R_1c_1c_1$' to '$(\exists x)R_1xx$' as follows: Suppose $R_1c_1c_1$. Then F_1c_1, by Synonomy. But then F_1*, by Existential Generalization.

Meaning postulates and synonomy rules also allow us to express assertions which would normally demand mixed strings of quantifiers—such as "$(x)(\exists y)R_1xy$". In this case we adopt the meaning postulate:

> If b is an individual constant, then R_1b* is equivalent to F_2b.

And we adopt the synonomy rule:

> If b is an individual constant, then from R_1b* infer F_2b —and vice-versa.

Now we can use '$\overline{\overline{F_2}}*$' to express that $(x)(\exists y)R_1xy$.

This strategy proved too cumbersome even for Weyl himself however. He eventually lets bound variables shoulder the burden of pronominal cross-reference, using, e.g., '$R_1xx \mid_{x\,=\,*}$' to express that $(\exists x)R_1xx$. Weyl does not give any reasons for preferring a variable-free logic to a more standard one—or vice-versa. He shifts from one approach to the other without any comment at all.

In Section 3, Weyl introduces some important semantic notions. He defines validity as truth under every interpretation—where an interpretation specifies a universe of discourse, assigns an object in this universe to each individual constant, and assigns an n-ary relation between objects of this universe to each n-ary relation symbol. Or at least this is what he does under one charitable and not too far-fetched reading of page 16. His talk of "recognition" and "self-evidence" is worrisome, but seems not really to involve him in a psychological account of logical validity. Weyl defines implication as validity of the conditional and equivalence as validity of the bi-conditional.

While Weyl's understanding of validity, implication, and equivalence is perhaps not subjectivistic, his views on the proper use of the axiomatic method most certainly are. An assertion is to be accepted as an axiom only if it is "recognized as true in immediate insight." And a *genuine* proof in an axiomatic system is one whose every step is approved by immediate insight. (More precisely: a genuine proof *for person P at time t* is one whose every step is approved at time t by P's immediate insight.) Weyl probably has something like Husserl's *Wesensschau* in mind here. In any case, readers who are not sure what "immediate insight" is supposed to be or who even doubt that there is such a thing are directed by Weyl to Husserl's *Logical Investigations* and *Ideas*.

Weyl introduces his set theory (more precisely, his theory of relations-in-extension) in a self-consciously Hegelian manner. He offers a series of distinct formulations—each one raising the theory to a "higher level." In Section 4, we take the first step on this journey to Absolute Knowledge. For our own presentation, let us adopt two classes of variables 'x_1', 'x_2', ... and 'X_1', 'X_2', ...–the former to be used when quantifying over individuals, the latter when quantifying over relations between individuals. We write our quantifier rules in such a way that there is no conflation of these two types. Let us also adopt an n+1-ary relation symbol ϵ_n for each natural number n. Now we can express Weyl's intitial comprehension scheme as follows (using familiar notation in place of Weyl's):

> If p is a sentence matrix in which neither any ϵ_n nor any relational variable occurs, then every closure of $(\exists X_1)(x_1) \ldots (x_n)(x_1 \ldots x_n \epsilon_n X_1 \longleftrightarrow p)$ is an axiom.

Weyl also proposes infinitely many extensionality axioms of the form $(X_1)(X_2) ((x_1) \ldots (x_n) (x_1 \ldots x_n \epsilon_n X_1 \longleftrightarrow x_1 \ldots x_n \epsilon_n X_2) \rightarrow X_1 = X_2)$. Weyl does not reduce relations-in-extension to sets. On the contrary, he makes set theory a part of the more general theory of relations-in-extension. Letting n = m in ϵ_n we obtain a theory of *m-ary relations-in-extension between individuals*. In particular, letting n = 1 we obtain a theory of *sets of individuals*. The above comprehension scheme is powerful enough to guarantee that there is, for example, an empty and a universal set of individuals. But it is not powerful enough to produce Russell's antinomy. That specter is banished both by the type-theoretical restrictions on our quantifier rules and by the exclusion of 'ϵ_1' from p. Given any finite number of individuals, we can prove that the unordered n-tuple of those individuals exists. And we are able to form the union and intersection of finitely many sets–as long as we are able to characterize each of those sets by a definite sentence matrix which contains neither any ϵ_n nor any relational variable. (We just form the disjunction or conjunction of those sentence matrices and take that as our p.)

However we are not able to prove that the union and intersection of any two (possibly uncharacterized) sets exist. (This would require us to use '$\in_1$' and relational variables in p.) Similarly, we can form the complement of a characterized set but not of an uncharacterized one. We are not, of course, able to form power sets.

The comprehension scheme we have just given does not allow us to prove that there are infinitely many individuals or that there is a set with infinitely many elements. It does, however, guarantee an infinite number of relations: at least one n-ary relation for each n. So Weyl might have tried to construct arithmetic within an expanded version of his theory of relations-in-extension, identifying the natural number n with, say, the universal n-ary relation. But Weyl thinks that a construction/reduction of the natural numbers within one's set or relation theory is of little significance as long as one's metatheory contains an arithmetical component which is simply *assumed*. And since Weyl believes that the natural numbers must be taken as primitives at *some* level, he does so at the very start: in the object language. In section 5, he treats the natural numbers as *sui generis* individuals and introduces an autonomous Dedekindian theory concerning them. He also introduces a form of recursive definition which we can express as follows. Let $p$xy be a binary sentence matrix in which neither any $\in_n$ nor any relational variable occurs. Then we are allowed to introduce a ternary relation symbol $\mathscr{R}$ via the postulate that $\mathscr{R}$xyn if and only if either

    a)   n = 1 and $p$xy

or

    b)   n = (m+1) and there is a z such that $\mathscr{R}$xzm and $p$zy.

For example, if we let $p$xy be '$x = y + 1$', then Rxyn if and only if $x = (y+n)$.

So far we have only introduced relations-in-extension which hold between *individuals*. But, as Weyl points out in Section 6, this will hardly suffice for a construction of analysis. We also require relations between relations between ... . So let us adopt an entirely new system of variables: 'x_1^0', 'x_2^0', ...; 'x_1^1', 'x_2^1', ...; 'x_1^2', 'x_2^2', ...; ... . A variable x_n^m is to be used when quantifying over objects of the mth level—where level 0 is that of individuals, level 1 that of relations between objects of level 0, level 2 that of relations between objects of either of the two previous levels, level 3 that of relations between objects of any of the three previous levels, and so on. In order to avoid the antinomies we bring our quantifier rules into line with this notion of cumulative levels. For example, if all x_1^3 have a certain property, then we should be able to infer that all x_1^2 have that property but *not* that all x_1^4 do. The comprehension scheme for level 1 relations corresponds to the one given above. For m $>$ 1 we have this:

> Suppose s, ..., t is a sequence of n natural numbers each less than m. And suppose ρ is a sentence matrix whose bound variables are all of level 0 and whose free variables other than x_1^s, ..., x_n^t are also all of level 0. Then every closure of $(\exists x_1^m) (x^s{}_1) \ldots (x_n^t) (x_1^s \ldots x_n^t \in_n x_1^m \longleftrightarrow \rho)$ is an axiom.

The most distinctive feature of this scheme is that it avoids impredicativity by restricting the variables of ρ (other than x_1^s, ..., x_n^t) to level 0—no matter what our choice of m is. An approach which is more in line with the ramified theory of types is to require merely that ρ's variables all be of a level less than m. This is sufficient to avoid any "definition of a relation in terms of a totality which contains that relation." But Weyl finds this alternative approach unnatural and inelegant because it disrupts the neat correlation between a relation's level and its extension and, therefore, breaks up the continuum into points which occupy distinct levels. More precisely, given this approach, there are relations x, y such that the lowest level which x occupies is different from the lowest level which y occupies even though, from an extensional point of view, x and y seem to be of the same species—even though, for example, the relata of x and y are all natural numbers. (I speak of *lowest* levels here because a relation of level n occupies all levels $> n$.) Surely it is not in the spirit of a theory of relations-in-extension to introduce irreducibly *intensional* distinctions such as these. One of Weyl's great achievements in *The Continuum* is his demonstration that one can avoid this undesirable situation without completely sacrificing constructive analysis.

Now that we have expanded our stock of comprehension axioms, we need to adopt new extensionality axioms to cover our larger universe. Given the suggested rendering of our quantifier rules, we can use $(x_1^m) \ldots (x_n^m)\, (x_1^m \ldots x_n^m \in_n x_1^{m+1} \longleftrightarrow x_1^m \ldots x_n^m \in_n x_{n+1}^p)$ to express the extensional equivalence of x_1^{m+1} and x_{n+1}^p when m+1 $\geq$ p and when it has been established that x_1^{m+1} is a non-empty n-ary relation (since otherwise the empty n-ary relation would be reckoned extensionally equivalent to every relation which happens not to be n-ary—indeed, every relation would turn out to be "extensionally equivalent" to every other relation). This last consideration forces us to add the clause $(\exists x_1^m) \ldots (\exists x_n^m)\, x_1^m \ldots x_n^m \in_n x_1^{m+1}$ to the antecedent of our extensionality scheme. Unfortunately, this means that our extensionality axioms fail to cover empty relations. We might want to get around this problem by introducing some device for restricting quantifiers to n-ary relations for any given n.

I have already noted that Weyl does not reduce relations-in-extension to sets. Neither does he reduce *functions* to relations-in-extension. He introduces functions more or less as follows. Let p be a sentence matrix whose only free variables are x_1^i, ..., x_m^j, x_{m+1}^k, ..., x_n^l and whose only bound variables are of level 0. Then we are allowed to introduce an (n-m)-ary functor f by means of the postulate (x_1^i) ... (x_m^j) (x_{m+1}^k) ... (x_n^l) $(x_1^1$ ... $x_m^j \in_m f x_{m+1}^k$ ... $x_n^l \longleftrightarrow p)$. The addition of such postulates strengthens our system significantly: they are certainly not just definitions. For example, as Weyl points out, we are now able to prove that every set of 0 level objects has a complement–regardless of whether the set in question has been characterized by a sentence matrix which contains variables only of level 0. We need only consider the postulate '(x_1^0) (x_1^1) $(x_1^0 \in_1$ Comp $(x_1^1) \longleftrightarrow x_1^0 \notin_1 x_1^1)$'. Similarly, we are now able to form the union and intersection of any finite number of possibly uncharacterized sets.

In Section 7, Weyl gives his system its final strengthening. First of all, he allows for the introduction of "0-ary" functors. That is, he allows us to introduce individual constants b via postulates of the form (x_1^u) ... (x_n^v) $(x_1^u$ ... $x_n^v \in_n b \longleftrightarrow p)$. Secondly, he allows both functors and individual constants to appear in the defining clauses (the p's) of the comprehension axioms and of the postulates which introduce functors. Finally, he establishes a new form of definition by recursion which exploits the functors which are now at hand. The simplest applications of this new technique are of the following sort. Let f be a unary functor which takes on n-ary relations as its values. Then we are allowed to introduce an (n + 2)-ary relation symbol $\mathcal{R}$ via the postulate that $\mathcal{R} x_1^i$ ... $x_{n+1}^k m$ if and only if either

    a) $m = 1$ and x_1^i ... $x_n^j \in_n f x_{n+1}^k$

or

    b) $m = p + 1$ and $\mathcal{R} x_1^i$ ... $f x_{n+1}^k p$.

As an example of how this form of recursive definition might be applied, Weyl introduces a cardinality operator for first level sets. This maneuver is really quite ingenious and merits some study. We begin by adopting two functors 'f' and 'g' via the postulates:

$$(x_1^0) \, (x_2^0) \, (x_1^1) \, (x_1^0 \in_1 f(x_2^0 x_1^1) \longleftrightarrow (x_1^0 \in_1 x_1^1 \, \& \, x_1^0 \neq x_2^0)).$$

$$(x_1^1)(x_1^2)(x_1^1 \in_1 g(x_1^2) \longleftrightarrow (\exists x_1^0)(x_1^0 \in_1 x_1^1 \, \& \, f(x_1^0 x_1^1) \in_1 x_1^2)).$$

$f(x_2^0 x_1^1)$ is the set of all elements of x_1^1 distinct from x_2^0. $g(x_1^2)$ is the set of all first-level sets which contain an element whose deletion produces a member of x_1^2. By recursion, we introduce a ternary relation symbol 'G' such that

$Gx_1^1 x_1^2 1$ if and only if $x_1^1 \in_1 g(x_1^2)$

$Gx_1^1 x_1^2 (n+1)$ if and only if $Gx_1^1 g(x_1^2) n$.

That is, $Gx_1^1 x_1^2 n$ if and only if x_1^1 contains n distinct elements whose deletion from x_1^1 produces an element of x_1^2. So we can use '$Gx_1^1 cn$' to express that x_1^1 contains at least n distinct elements—where the individual constant 'c' is introduced by means of the postulate '$(x_1^1)(x_1^1 \in_1 c \longleftrightarrow (P \vee -P))$', i.e. where c is the set of all first-level sets. We introduce the functor 'Card' via the postulate that $(x_1^1) \, (n \in_1 \mathrm{Card} \, (x_1^1) \longleftrightarrow Gx_1^1 cn)$.

The reader may think that this definition of 'Card' is unnecessarily complicated and indirect. But Weyl's twistings and turnings have a purpose. Notice that the right-hand sides of the postulates Weyl has used to introduce 'f', 'g', and 'c' contain no quantifications over objects of any level higher than 0. Weyl had to stray from the most direct path in order to avoid forbidden higher level quantifications. Notice also that the sentence

Preface

It is not the purpose of this work to cover the "firm rock" on which the house of analysis is founded with a fake wooden structure of formalism—a structure which can fool the reader and, ultimately, the author into believing that it is the true foundation. Rather, I shall show that this house is to a large degree built on sand. I believe that I can replace this shifting foundation with pillars of enduring strength. They will not, however, support everything which today is generally considered to be securely grounded. I give up the rest, since I see no other possibility.

At the center of my reflections stands the conceptual problem posed by the continuum—a problem which ought to bear the name of *Pythagoras* and which we currently attempt to solve by means of the arithmetical theory of irrational numbers. The main ideas are developed in Chapter 1, deliberately in such a way that this part forms a self-contained whole. In this first chapter, I present the principles with the help of which, in Chapter 2, I systematically begin the construction of analysis and carry through its initial stages. It was unavoidable that in the second chapter some things which had already been stated often—in rather varied terminological garb—had to be repeated. This was done as sparingly as is possible without compromising the integrity of the picture being sketched. Nonetheless, I would like to be understood not merely by professional academics, but also by all students who have become acquainted with the currently canonical and allegedly "rigorous" foundations of analysis.

Foundational research has not yet progressed to the point at which one author can build on the results of another. So it is not

1

a good idea to interrupt the systematic presentation of one's own thoughts by references to the position of other researchers on the same questions and by arguments with them. Thus I have chosen to do this only in the concluding remarks of Chapter 1.

Although this is primarily a mathematical treatise, I did not avoid *philosophical* questions and did not attempt to dispose of them by means of that crude and superficial amalgamation of empiricism and formalism which still enjoys considerable prestige among mathematicians (even though it is attacked with gratifying clarity in Frege [1893]). Concerning the epistemological side of logic, I agree with the conceptions which underlie Husserl (1913a). The reader should also consult the deepened presentation in Husserl (1913b) which places the logical within the framework of a comprehensive philosophy. Our examination of the continuum problem contributes to critical epistemology's investigation into the relations between what is immediately (intuitively) given and the formal (mathematical) concepts through which we seek to construct the given in geometry and physics.

Hermann Weyl

Zurich, November 1917

Preface to the 1932 Reprint

The point of view adopted in this monograph continues to strike me as a natural transitional stage in the development of foundational research. However, in the period since its appearance, my work has been superseded by two trends identified by the catchwords Intuitionism and Formalism. Still, this deeper grounding of the foundation has not led to an even moderately satisfying or defensible conclusion; things remain in a state of flux. It seems not to be out of the question that the limitations prescribed in the present treatise—i.e., unrestricted application of the concepts "existence" and "universality" to the natural numbers, but not to sequences of natural numbers—can once again be of fundamental significance. It would not be possible, without radical rebuilding, to bring the content of this monograph into harmony with my current beliefs*—and such a project would keep me from satisfying other demands on my time. So an unrevised reprint has been arranged.

For an overview of the systematic form in which analysis must be erected starting from the standpoint of this treatise, the reader should consult Weyl (1919).**

Göttingen, June 1932

*See Weyl, 1921, 1924, 1925, 1928, 1929, 1931. (Trans.)
**See the Appendix of this volume for a translation of this paper. (Trans.)

Chapter 1

Set and Function
Analysis of Mathematical Concept Formation

Logical Section

§1. PROPERTY, RELATION, EXISTENCE

A *judgment* affirms a *state of affairs*. If this state of affairs obtains, then the judgment is *true*; otherwise, it is untrue. States of affairs involving *properties* are particularly important. (Indeed, logicians often made the mistake of ignoring every other sort of state of affairs.) A judgment involving properties asserts that a certain object possesses a certain property, as in the example: "This leaf (given to me in a present act of perception) has this definite green color (given to me in this very perception." A property is always affiliated with a definite category of object in such a way that the *proposition* "*a* has that property" is *meaningful*, i.e., expresses a judgment and thereby affirms a state of affairs, only if *a* is an object of that category. For example, the property "green" is affiliated with the category "visible thing." So the proposition that, say, an ethical value is green is neither true nor false but meaningless. A judgment corresponds only to a *meaningful* proposition, a state of affairs only to a *true* judgment; a state of affairs, however, *obtains*—purely and simply. Perhaps meaningless propositions can appear only in thought about language, never in thought about things. At any rate, a great danger of language is that it makes possible meaningless combinations of expressions [such as "honesty" and "green"] which signify basic components of judgments; and such meaningless combinations can have the same grammatical form as proposi-

5

tions which express genuine judgments.[1] For even if a proposition's grammatical structure fails to reveal that it is meaningless (as in the case of a proposition having the form "The object *a* has the property *P*"), it does not follow that the proposition is meaningful—just as little does a judgment which is not logically *absurd* (which is not recognized as untrue independently of its material content, purely on the grounds of its "logical" structure) thereby have to be *true* (cf. §3). But if a proposition "*a* has the property *P*" expresses a judgment, the same holds for the corresponding negation "*a* does not have the property *P*", and formal logic is then entirely correct when it claims that, of these two judgments, it is always the case that one is true, the other false.

Propositions which embody a judgment concerning a property (and *only* these) have the well known {subject–copula–predicate} structure. But anyone who forgets that a proposition with such a structure can be meaningless is in danger of becoming trapped in absurdity—as a famous "paradox," essentially due to Russell, shows.[2] Let a word which signifies a property be called *autological* if this word itself possesses the property which it signifies; if it does not possess that property, let it be called *heterological*. For example, the German word "kurz" (meaning "short") is itself *kurz* (i.e., is itself short—for a word in the German language which consists of only four letters will without question have to be described as a short one[3]); hence "kurz" is autological. The word "long," on the other hand, is not itself long and, so, is heterological. Now what about the word "heterological" itself? If it is autological, then it has the property which it expresses and, so, is heterological. If, on the other hand, it is heterological, then it does not have this property and, so, is autological. Formalism regards this as an insoluble contradiction; but in reality this is a matter of scholasticism of the worst sort: for the slightest consideration shows that absolutely no sense can be attached to the question of whether the word "heterological" is itself auto- or heterological.

We cannot set out here in search of a definitive elucidation of what it is to be a state of affairs, a judgment, an object, or a property. This task leads into metaphysical depths. And concerning it one must consult men, such as Fichte, whose names may not be mentioned among mathematicians without eliciting an indulgent smile.

In addition to judgments involving properties, those involving *relations* are of importance to us. Examples: "That man over there is one of my uncles." "Point A lies between B and C." "The number 5 is the successor of 4." At this juncture, we should note a similarity between these relational propositions and those involving properties. We can associate the *judgment scheme* "x is the successor of y," which contains the "variables" x, y, with the relation mentioned in the third example. A definite judgment arises from this scheme when we substitute any two numbers for the variables; thus we assert that, e.g., this judgment is true "for" $x = 5$, $y = 4$. Every variable, every "blank," of the judgment scheme is affiliated with a definite category of object (in our example, with the category "number"). And only if the blank is filled by an object of this category does the scheme yield a meaningful proposition concerning which the question of whether it is true or not then arises. For the sake of simplicity, I am speaking here only of relations such that every blank of each of the corresponding judgment schemes is connected with the *same* category of object. Contrary to the prevailing terminological conventions of mathematics, it must be stressed that the propositions "5 is the successor of 4" and "4 is the predecessor of 5" express *one and the same* relation between 4 and 5 and that it is therefore improper to speak of two distinct relations, of which one is the "inverse" of the other. The corresponding judgment scheme contains two blanks (which, naturally, do not "enjoy equal status"). By placing these blanks first in the one and then in the other *order of succession* (as is required for linguistic symbolization), we obtain each of the propositions just mentioned. But, obviously, such an order of succession is not intrinsic to the relational state of affairs.[4]

Let a definite category of object (e.g., "spatial point") be "immediately given" (i.e., exhibited in intuition). In what follows, we shall restrict our attention to this category. So let us assume that we have been presented with certain individual properties and relations (R) which apply to objects of our chosen category and which (like the relation "lies between") are connected (by every blank of their judgment schemes) with this category. We should not concentrate solely on those judgments (some true, others false) which concern properties and relations and which arise when judgment schemes corresponding to members of (R) are filled by immediately exhibited objects of our category. For *existential judgments* play an essential role in mathematics. The concept of existence is overburdened by metaphysical enigmas. Luckily, however, for our current purposes we need only assume the following. If, say, $P(x)$, $P'(x)$, $R(xy)$ are among the judgment schemes (R) (letting x,y signify blanks) and a is an individual given object [of our category], then propositions such as "*There is* an object (of our category) of which both $P(x)$ and $P'(x)$ are true (i.e, which has both the properties P and P')" or "*There are* objects x which stand to a in the relation $R(xa)$" are said to be meaningful—that is, they affirm definite (existential) states of affairs concerning which the question of whether they obtain or not can now be raised.[5] It is in this sense that we understand the hypothesis that *the characteristic features of the categorical essence under consideration* are supposed to determine *a complete system of definite self-existent objects* [namely, the *extension* of that category]. Our remarks are

$\cdot N(\gamma z)$. Thus a new complex judgment arises from two judgments F and N when, first, the blanks of the one are made to coincide in part with those of the other and, second, the and–connection is established. The manner in which the blanks of the two original judgments are supposed to coincide can, as our example indicates, be expressed in the symbolism by marking the coinciding blanks with the same letter. In general, the and–connection allows not just one, but several new judgments to be produced from two given ones, depending on how the blanks of the one and the other are made to coincide: that is, either not at all or partly or completely. Of course, such and–connections of a judgment with *itself* are also possible; for example, $N(xy) \cdot N(\gamma x)$ ("x is a nephew of y and, at the same time, y is a nephew of x").

4. The connection by "or," for which we use the symbol $+$, complements the connection by "and." Here are some examples of its use: $P(x) + P'(x)$ means "x is either red or spherical"; $F(xy) + N(\gamma z)$ means "Either x is the father of y or y is a nephew of z"; and $N(xy) + N(\gamma x)$ means "Either x is a nephew of y or y is a nephew of x," that is, "x is either a nephew or an uncle of y."[7] Our remarks above about how blanks of the one judgment are to coincide with those of the other apply here too.[8]

5. If, e.g., $U(xyz)$ is a judgment with three blanks and a is a given object of our category, then the judgment $U(xya)$, which is produced by the operation of "*filling in*," is one with only two blanks. In particular, a *closed* judgment (i.e., one without blanks), a judgment in the proper sense, which affirms a state of affairs, arises from a judgment scheme when all its blanks are filled by certain given objects of our category.

6. Let us continue to suppose (by way of example) that $U(xyz)$ is a judgment with three blanks. Then we can form the judgment $U(xy*) = V(xy)$, which means "*There is* an object z (of our category) such that the relation $U(xyz)$ obtains."[9] Similarly, we can form $U(*y*)$, meaning "*There is* an object x and an object z such that $U(xyz)$ is true."[10] The number of blanks of a judgment scheme will be reduced by application of this principle too. If no blanks at all are left, then here again a judgment in the proper sense arises, about which it is then appropriate to ask whether it is true or not. For example, if $F(xy)$ means "x is the father of y" and, accordingly, $F(Iy)$ means "I am the father of y," then $F(I*)$ means "There is someone of whom I am the father," that is, "I am a father."

With regard to 5 and 6, observe that, e.g., if a signifies a given object, then from $U(xy) \cdot V(xy) = W(xy)$ there follows $U(xa) \cdot V(xa) = W(xa)$, but certainly not $U(x*) \cdot V(x*) = W(x*)$; rather $U(x*) \cdot V(x*) = T(x**)$ follows, if one introduces $U(xy) \cdot V(xz) = T(xyz)$.[11] Further, it is of interest to note that principle 3 is reducible to 4 with the help of 1, that is, negation (see §3).

By applying principles 1 through 6 to simple judgment schemes, we obtain new ones. We can once again apply the same principles to these new judgment schemes and to the primitive ones and, thereby, again obtain new judgment schemes—and so on, with as many repetitions and in as many combinations as one wishes. Among the endless abundance of judgment schemes which arise in this way, those which possess *one* blank are the judgment schemes of *derived properties*; those which possess two or more blanks are the judgment schemes of *derived relations*. But those which possess no blanks at all and which, therefore, are judgments in the proper sense and, so, affirm a state of affairs,[12] we call the *pertinent judgments* of our discipline. If we knew, concerning each of these pertinent judgments, whether it were true or not, we would possess a complete knowledge of the objects of the underlying category with respect to the immediately exhibited properties and relations from which we started. Our principles define the logical function of the concepts "not," "and," "or," and "there is" in an exact way. With respect to their logical form, it is impossible to divide the pertinent judgments into those dealing with properties or with relations or with existence, or into affirmative and negative judgments, or into any other traditional classifications. Rather such a judgment, in general, has a very complex *logical structure* which can be completely described only by indicating in what manner and sequence, and by means of which combinations, that judgment arose from the underlying simple judgment schemes by application of our six principles. Here we are infinitely far removed from the old doctrine that a proposition always consists of subject, predicate, and copula.

We are going to consider a few examples of the combined application of our principles. But first we note that we must use a combination of 1 and 6 (negation and "there is") in place of "all," which expresses universality. "Every object has such and such a property" means "There is no object which lacks the relevant property." Judgments of the following form occur frequently in mathematics (letting $U(xy)$ signify the judgment scheme of a relation with two blanks x,y): "For every x there is a y such that $U(xy)$ obtains." We can construct this judgment as follows. From $U(xy)$ we form $U(x*) = A(x)$; from the later we form the negation $\overline{A}(x) = B(x)$; and from this we form $B(*)$ and its negation $\overline{B(*)}$ (not to be confused with $\overline{B}(*)$, i.e., $A(*)!$[13]): this is the statement mentioned just above (which, naturally, contains no "blanks").

Example A. Domain of Objects: Points of Plane Geometry.

Let $D(xyz)$ mean that x and y are equidistant from z. The propositions "xyz lie on a line" or, alternatively, "The relation $L\,(xyz)$ holds" mean that there are two distinct points p and q such that p and q are equidistant from x, y, and z. How might we construct this relation L? First, by principles 3 and 1 and with the help of the identity scheme J, we form

$$D(pqx) \cdot D(pqy) \cdot D(pqz) \cdot \bar{J}(pq) = F(xyzpq).$$

Then we can form $F(xyz**) = L(xyz)$.

Example B. Domain of Objects: Real Numbers.

Let $f(x)$ be a function of the real argument x. We want to analyze the judgment "f is uniformly continuous." According to the standard explanation, this means that for every positive number ε there is a positive number δ such that for any two numbers x and y which satisfy the inequality $|\,x\text{-}y\,| < \delta$, the inequality $|\,f(x)\text{-}f(y)\,| < \varepsilon$ always holds too. Let

$$A\,(xy\varepsilon) \text{ signify the relation } |\,x{-}y\,| < \varepsilon.$$

$$F(xy\varepsilon) \text{ signify the relation } |\,f(x){-}f(y)\,| < \varepsilon.$$

And let $P(\varepsilon)$ mean that ε is positive. Using 1 and 3, we now form

$$A(xy\delta) \cdot \bar{F}(xy\varepsilon) = \mathrm{B}(xy\varepsilon\delta);$$

and from this we form

$$B(**\varepsilon\delta) = \mathrm{C}(\varepsilon\delta) \text{ and its negation } \overline{C}(\varepsilon\delta);$$

from this

$$\overline{C}(\varepsilon\delta)\cdot P(\delta) = Q(\varepsilon\delta) \text{ and } Q(\varepsilon*) = R(\varepsilon);$$

from the negation of R we form $\overline{R}(\varepsilon)\cdot P(\varepsilon) = S(\varepsilon)$; and from this we form the closed judgment $U = S(*)$. "f is uniformly continuous" means that the negation $\overline{U}$ of the above judgment is true.

Example C.

Let us also explain the proposition "f is continuous at every value of its argument." Let $B(xy\varepsilon\delta)$ be as above. Then we form, in turn,

$$B(x*\varepsilon\delta) = C(x\varepsilon\delta); \text{ its negation } \overline{C}(x\ \varepsilon\delta);$$

$$\overline{C}(x\ \varepsilon\delta)\cdot P(\delta) = Q(x\ \varepsilon\delta); Q(x\varepsilon*) = R(x\varepsilon);$$

and from $\overline{R}(x\varepsilon)\cdot P(\varepsilon) = S(x\varepsilon)$ we form $S(**) = V$. The negation $\overline{V}$ of V is the statement we wanted.

The symbolism we are using is, as our examples indicate, unwieldy; but that need not concern us. On the other hand, the list of the six principles of definition is, just by itself (unless we are wrong to think it *complete*), of considerable importance for logic.

§3. LOGICAL INFERENCE. AXIOMATIC METHOD

We call a judgment [14] *general* if Pr. 5 is not employed in its construction, so that the blanks are filled only by "*", i.e., "there is." In mathematics, we deal exclusively with such judgments—which it would also be reasonable to call "existential judgments." But if Pr. 5 is employed and individual, immediately exhibited objects of our category thereby enter into the judgment, we speak of a *particular* judgment.[15] If $P(x)$ is a judgment scheme with one blank which arises from the primitive properties and relations through application of any of our principles *except* 5 and if there is one and only one object $x=a$ such that $P(x)$ holds, then a is called an *individual* (which can be characterized by means of its properties). Naturally, the exclusion of Pr. 5 is essential here; for, otherwise, if we again take J to be identity, the judgment scheme $J(xa)$ with the blank x would designate a property, "being a," which applies only to the object a, and the concept "individual" would be vacuous.

The arithmetic of natural numbers supplies us with an example of a domain of individuals. The sole fundamental relation which underlies this discipline is $S(n\ n')$, which holds if n' is the number immediately succeeding n. 1 is characterized by the property "I": that is, there is no number of which 1 is the successor (the number sequence begins with 1); i.e.,

$$\overline{S(*x)} = \mathrm{I}(x).$$

It is simply a *fact* that there is one and only one number with this property I; we happen to *call* it 1. Now 2 may be characterized by the Property II, namely the property of immediately succeeding the number 1 which we just defined:

$$\mathrm{I}(\gamma)\cdot\ S(\gamma x)\ =\ S_2(\gamma x);\ S_2(*x)\ =\ \mathrm{II}(x).$$

We can characterize 3, 4, etc. in analogous ways. Clearly, every number is an individual. The propositon $1 + 2 = 3$ embodies a particular judgment if 1, 2, 3 are immediately exhibited numbers. But, in fact, it is impossible for a number to be given otherwise than through its position in the number sequence,[16] i.e., by indicating its characteristic property. So let us interpret the proposition $1 + 2 = 3$ as follows: "There are three number $x,\ y,\ z$ of which $\mathrm{I}(x)$, $\mathrm{II}(y)$, $\mathrm{III}(z)$ and $x + y = z$ are true." We see then that $1 + 2 = 3$ actually embodies a "general" judgment. In some cases, for example in arithmetic, all objects of the category under consideration are "individuals" (in the sense defined above). Diametrically opposed to this is the situation wherein each judgment scheme $P(x)$ which contains a single blank and which arises from the basic properties and relations without any application of Pr. 5 is always true either of *each* or of *no* object. In this situation, we may call our category *homogeneous* (with respect to these basic properties and relations). For example, the spatial points of Euclidean geometry are of this sort and this is precisely why we call geometrical space *homogeneous*.[17]

Some pertinent judgments we recognize as true purely on the basis of their logical structure—without regard either to the characteristics of the category of object involved or to the extensions of the basic underlying properties and relations or to the objects used in the operation of "filling in" when Pr. 5 is applied. Such judgments which are true purely on account of their formal (logical) structure (and which thus possess no "material content") we wish to call (*logically*) *self-evident*. A judgment whose negation is self-evident is called *absurd*. If $U \cdot \overline{V}$ is absurd, then the judgment V is a *logical consequence* of U; if U is true, then we can be certain that V is also true. If V is a logical consequence of U and, conversely, U is a logical consequence of V, then the two judgments U and V are *equivalent*. One of the most important tasks of logic (the doctrine of inference) is to describe completely those propositional structures which guarantee the self-evidence of judgments. Logic assembles certain "elementary" structures of this kind, from which all such propositional structures arise by means of a process of "composition" which remains to be characterized more precisely. We shall not try to decide whether this task has been completed in an entirely satisfactory way by either the traditional or the so-called mathematical logic. Instead, we merely call to mind some examples.

Taking U to be any judgment at all, $U + \overline{U}$ is self-evident and $U{\cdot}\overline{U}$ is absurd. The judgments U and $\overline{\overline{U}}$ are equivalent. Given two judgments U and V, $U{\cdot}V$ is equivalent to $\overline{\overline{U}{+}\overline{V}}$. If $U(x)$, $V(x)$, $W(x)$ are any three judgment schemes with one blank, then the rule of the syllogism reads:

$$\overline{U{\cdot}\overline{V}}(*) \cdot \overline{V{\cdot}\overline{W}}(*) \cdot (U{\cdot}\overline{W}(*)) \text{ is absurd.}[18]$$

$\overline{U \cdot \overline{V}}(*)$ means that there is no object x such that $U(x)$ is true but $V(x)$ is not; i.e. V is a property of all objects which possess the property U.

The well known significance of logical inference for science goes without saying. And everyone knows exactly what role the deductive method plays in mathematics. The states of affairs with which mathematics deals are, except for the very simplest ones, so complicated that it is practically impossible to bring them into full givenness in consciousness and, in this way, to grasp them completely. So the situation in mathematics is actually as follows. Mathematics concerns itself with *pertinent, general, true* judgments. Among these are a few which are immediately recognized as true, the *axioms*, say U_1, U_2, U_3, U_4, which are such that all other pertinent, general, true judgments are logical consequences of these few, i.e. of $U_1 \cdot U_2 \cdot U_3 \cdot U_4$. As our recent discussion of the laws of logic suggests, a system of "elementary" inferences which, in general, takes the form of a multi-limbed organism is needed in order to demonstrate that a judgment U is a consequence of the axioms. In order to be communicated, this system must then be transformed in an artificial way into a chain of interlocking links. It is in this manner that mathematical *proof* takes place. Here all the insight which is to be effected is concentrated on the logical inferences and is no longer directed toward the objects and states of affairs about which determinations have been made.[19] (It need hardly be easily adapted to more complicated cases in which, from the beginning, there are several underlying categories of object rather than just one (as, e.g., in the geometry of Euclid where there are three categories: point, line, plane).

§2. THE PRINCIPLES OF THE COMBINATION OF JUDGMENTS

By *simple* or *primitive* judgment schemes (or even, more briefly, "simple judgments"–where, for the moment, we take the word "judgment" in a wider sense than we have so far) we mean those which correspond to the individual immediately given properties and relations. To these we add the *identity* scheme $J(xy)$ (meaning "x is identical to y"–i.e., "$x=y$"). *Complex* judgment schemes are to be obtained from these simple ones by applying the following principles.[6]

1. From the judgment scheme U we obtain its negation $\overline{U}$. For example, if we let $U(xy)$ mean that x is the successor of y, then $\overline{U}(xy)$ means that x is not the successor of y.

2. In a judgment scheme with several blanks, some of these blanks can be made to coincide, can be "identified," with one another. This yields a new judgment scheme. For example, from the relational scheme $N(xy)$, meaning "x is a nephew of y," we obtain $N(xx)$, meaning "x is his own nephew."

3. Two judgments can be combined with one another by an "and," as in the following examples. If $P(x)$ means "x is red" and $P'(x)$ means "x is spherical," then we designate the complex judgment "x is both red and spherical" by $P(x) \cdot P'(x)$. And if $F(xy)$ means "x is the father of y" and $N(xy)$ means "x is a nephew of y," then the complex judgment "x is the father of y and y is a nephew of z," which asserts a relation between three persons x, y, z, is to be designated by $F(xy)$

mentioned that the discovery of mathematical truths and the subsequent grasp of them by the understanding occurs much more "substantively" and much less "formally." Here we are merely discussing the systematic presentation.) However, we must stress that it is merely a scientific *belief* that, for example, all pertinent, general, true judgments about points, lines, and planes are derivable from the geometrical axioms. We are unable to grasp in genuine *insight* that this is so or even to "prove" it in a logical way on the basis of the logical laws themselves. Were this to happen one day, then this insight would lead to a definite, methodical proof procedure for reaching (in finitely many steps) a decision concerning the truth or falsity of every (pertinent, general) geometrical judgment. Thus mathematics would, in principle, be *trivialized.*[20]

Today it is frequently maintained that axioms are *stipulations* and that, for example, Fermat's proposition ("there are no whole numbers $x \neq 0$, $y \neq 0$, $z \neq 0$, $n > 2$ such that $x^n + y^n = z^n$") asserts merely that this judgment is a *consequence* of the arithmetical axioms. Thus the axioms define, so to speak, the meaning of "there is." (That exists whose existence can be derived from the axioms.) But quite apart from the fact that such a "hypothetico-deductive game" is worthless (if the axioms are devoid of any meaning which has cognitive significance), this view is even logically untenable. An example: If we define the irrational numbers in Dedekind's manner, then our definition immediately determines when a rational number is to be called *smaller* ($<$) than a real number. If a, b are any two real numbers—each of which is assumed to be defined as an individual by a property which applies to it and only to it—then $a < b$ if there is a rational number r such that $a < r$ and $r < b$. If "there is" is interpreted here in accordance with the above-mentioned view, then $b \leq a$ holds only if the axioms imply that there is no rational number r which satisfies the inequalities

$a < r$, $r < b$. Thus the judgments $a < b$ and $b \leq a$ do not amount to exhaustive alternatives, since it can very likely happen that neither the existence nor the non-existence of such a rational number r is a consequence of the arithmetical axioms. The interpretation under consideration proves to be feasible only when one knows that the axioms are *consistent* and *complete* in the sense that of two "antinomous" pertinent judgments U and $\overline{U}$ always one and only one is a logical consequence of the axioms. But we do not *know* this (although we may believe it). And if this belief is one day to be transformed into insight, then, clearly, since logical inference consists of iterating certain elementary logical inferences, we will attain this insight only through our intuition of iteration, i.e., of the infinite repetition of a procedure. But from this intuition we also directly obtain the fundamental arithmetical insights into the natural numbers on the basis of which the whole *mathesis pura* is logically constructed. [And this undercuts the claim that arithmetical axioms are mere stipulations.]

Mathematical Section

§4. SETS

Finite sets can be described in two ways: either in *individual* terms, by exhibiting each of their elements, or in *general* terms, on the basis of a rule, i.e., by indicating properties which apply to the elements of the set and to no other objects. In the case of infinite sets, the first way is impossible (and this is the very essence of the infinite). The "characteristic properties" which are appropriate for the general description of infinite sets are the primitive properties and relations and those derived from them in the manner of §2; these constitute the sphere of the "attributable" properties, Hence:

To every primitive or derived property P there corresponds a **set** (P). *The expressions "An object a has the property P" (or "The relevant judgment scheme P(x) containing one blank is true for x=a") and "a is an element of the set (P)" have the same significance. The same set corresponds to two such properties P and P′ if and only if every object (of our category) which has the property P also has the property P′, and conversely.*

Therefore, how two *sets* (in contrast to *properties*) are defined (on the basis of the primitive properties and relations and individual objects exhibited by means of the principles of §2) does not determine their identity. Rather an *objective* fact, which is not decipherable from the definition in a purely logical way, is decisive; namely, whether each element of the one set is also an element of the other, and conversely. Moreover, we see that the description of a finite set in individual terms is, considered formally, just a special case of that based on a rule. For example, if *a,b,c* are three objects of our category, then

$$P(x) = J(xa) + J(xb) + J(xc)$$

is the judgment scheme of the derived property "being *a* or *b* or *c*"; and the set having just those three objects as its elements corresponds to this property.

As for relations and their judgment schemes, we can adopt the same view: namely, that the objective range of applicability (i.e., the extension) rather than the form of definition (i.e., the intension or meaning) determines their identity. Just as a *set* corresponds to each property, so a *functional connection* corresponds to each relation. This expression "functional connection" (for which I find no shorter or more appropriate alternative) is supposed to remind us from the start that here we are at the root of the mathematical concepts "function, assignment, mapping." But instead of using this expression, one may, depending on the number of blanks, speak of *2-, 3-, 4-dimensional sets*. What we above called a set must then be characterized more precisely as a "one-dimensional set." The rest of our terminology is governed by our choice of one or the other term. For example, if *a, b* stand to one another in the binary relation *R*, then we will say either that *a, b* form a system of elements of the corresponding two-dimensional set (*R*) or that they satisfy the functional connection (*R*).

A very important situation involving multi-dimensional sets remains to be considered. Let $U(xy)$ and $V(xy)$ be any two binary judgment schemes. If there are no objects $x=a$, $y=b$ of our category such that $U(ab)$ obtains but $V(ab)$ does not or $V(ab)$ obtains but $U(ab)$ does not, then the same functional connection corresponds to these two relations. Now the claim we are making here about the two judgment schemes *U* and *V* does not explicitly mention any intrinsic relation at all which holds between them, though it does clearly presuppose that their blanks can be made to coincide completely with one another in a certain way. However, if a functional connection is to correspond to the judgment scheme $U(xy)$ in a way which satisfies our claim, we must assume that the blanks of the judgment scheme are arranged in a definite order of succession. We then wish to call the judgment scheme *subject-ordered*. This arrangement is indicated in the symbolism simply by the order of succession (from left to right, as in our normal writing) of the letters which represent the blanks. Our claim has a clear sense with respect to *subject-ordered* binary relations only when it is understood that the objects *a* and *b* are to be inserted in the ordered blanks of the two judgment schemes in the *same* order of succession.

To every subject-ordered judgment scheme of a primitive or derived relation there corresponds a **functional connection**, *a* **multi-dimensional set***; the same functional connection corresponds to two such judgment schemes (with the same number of blanks) if and only if whatever objects of our category satisfy the one relation always satisfy the other as well in the same order of succession, and conversely.*

The principles of §2 now change into ones governing the "production" of one- and multi-dimensional sets. In the domain of sets, the operation of forming the *complement* corresponds to negation (Pr. 1). The *intersection* and *sum* of two sets arise in accordance with Pr. 3 and 4. If we use Pr. 5 to insert a given object *a* into, say, a ternary relation $U(xyz)$, then the "*cross-section*" $z=a$, which is a two-dimensional set, arises from the corresponding three-dimensional set. Mimicking the terminology of analytic geometry, we may call Pr. 6 the principle of *projection*.

The one- and multi-dimensional sets form a new derived system of ideal objects over and above the given primitive domain of objects. The new system arises from the original one by means of what I call the *mathematical process*. In fact, I believe that this process of concept formation reveals the characteristic feature of the mathematical way of thinking. Obviously, these new objects, the sets, are altogether different from the primitive objects; they belong to an entirely separate sphere of existence.

No one can describe an infinite set other than by indicating properties which are characteristic of the elements of the set. And no one can establish a correspondence among infinitely many things without indicating a *rule,* i.e., a relation, which connects the corresponding objects with one another. The notion that an infinite set is a "gathering" brought together by infinitely many individual arbitrary acts of selection, assembled and then surveyed as a whole by consciousness, is nonsensical; "inexhaustibility" is essential to the infinite. So our view is that the transition from the "property" to the "set" (of those things which have the property) signifies merely that one brings to bear the *objective* rather than the purely logical point of view, i.e., one regards *objective* correspondence (that is, "relation in extension" as logicians say) established entirely on the basis of acquaintance with the relevant objects as decisive rather than logical equivalence. Therefore I contrast the concept of set and function formulated here in an exact way with the *completely vague* concept of function which has become canonical in analysis since Dirichlet and, together with it, the prevailing concept of set. Although some of the principal components of the mathematical edifice, such as elementary geometry, arithmetic and rational algebra, are in good condition, the same cannot be said of analysis and set theory (cf. especially §6). Everyone agrees that the much lauded critique to which the nineteenth century subjected the foundations of classical analysis was justified. And, certainly, this critique is responsible for an immense advance in the rigor of thought. But what was positively erected in place of the old is, *if one's glance is directed to the ultimate principles*, even more unclear and assailable than what it replaced–although it is certain that most of the achievements of modern critical research can be used anew as building material for a definitive founding of analysis. But as things now stand we must point out that, in spite of Dedekind, Cantor, and Weierstrass, the great task which has been facing us since the Pythagorean discovery of the irrationals remains today as unfinished as ever; that is, the *continuity* given to us immediately by intuition (in the flow of time and in motion) has yet to be grasped mathematically as a totality of discrete "stages" in accordance with that part of its content which can be conceptualized in an "exact" way. More or less arbitrarily axiomatized systems (be they ever so "elegant" and "fruitful") cannot further help us here. We must try to attain a solution which is based on objective insight. At this point, we would do well to explore somewhat further the consequences for the foundations of analysis and set theory of our view concerning the concepts of set and function.

For the moment, we shall confine ourselves to some general remarks on this topic: We must proceed from a definite sphere of operation. And the existence of a set or correspondence is determined by the objective connections which obtain between the objects of the given categories and which can be expressed by means of the underlying basic properties and relations. Thus, contrary to Cantor's proposal, no universal *scale of infinite cardinal and ordinal numbers* applicable to every sphere of operation can exist. (This does not, however, rule out a universal set theory.) The gap between the finite and the infinite, which set theory seemed to close, again opens wide before us. A set-theoretic treatment of the natural numbers such as that offered in Dedekind (1888) may indeed contribute to the systematization of mathematics;[21] but it must not be allowed to obscure the fact that our grasp of the basic concepts of set theory depends on a prior intuition of iteration and of the sequence of natural numbers.

§5. THE NATURAL NUMBERS: RICHARD'S ANTINOMY.

The remarks made in the preceding sections can, in particular, be applied to the category consisting of the ideal objects known as *natural numbers*. A single basic relation, whose meaning is immediately exhibited, underlies this category—namely, the relation $S(xy)$ which holds between two natural numbers x, y when y is the immediate successor of x. The following simple facts about S obtain: For every number x there is one and only one y such that $S(xy)$ holds. There is a unique number 1 which is not the immediate successor of any number; but every other number is the immediate successor of exactly one number. Finally, the important inferential technique of *complete induction* is based on the circumstance that one can eventually reach any number whatever by starting from 1 and proceeding from each number to its successor.

It is characteristic of every mathematical discipline that 1) it is based on a sphere of operation such as we have presupposed here from the beginning; that 2) the natural numbers along with the relation S which connects them are always associated with this sphere; and that 3) over and above this composite sphere of operation, a realm of new ideal objects, of sets and functional connections is erected by means of the mathematical process [discussed above, page 22] which may, if necessary, be repeated arbitrarily often. The old explanation of mathematics as the doctrine of number and space has, in view of the more recent development of our science, been judged to be too narrow. But, clearly, even in such disciplines as pure geometry, analysis situs, group theory, and so on, the natural numbers are, from the start, related to the objects under consideration. So from now on we shall assume that at least one category of object underlies our investigation and that at least one of these underlying categories is that of the natural numbers. If there is more than one such category, we should recall the observation in §1 that each blank of a judgment scheme of a primitive or derived relation is affiliated with its own definite category of object. If the underlying sphere of operation described at the beginning of this paragraph is that of the natural numbers, without anything further being added, then we arrive at *pure number theory*, which forms the centerpiece of mathematics; its concepts and results are clearly of significance for *every* mathematical discipline.

If the natural numbers belong to the sphere of operation, then a new, important, and specifically mathematical principle of definition joins those enumerated in §2; namely, the principle of *iteration* (definition by complete induction) by virtue of which the natural numbers first come into contact with the objects of the remaining categories of the underlying sphere of operation (if there are any). For example, in pure number theory, we need this principle in order to manufacture the fundamental arithmetical relations

$$m \, < \, n, \; m + n = p, \; m \cdot n = p$$

taking S as our starting point (cf. Chapter 2, §1). In the foundations of geometry, iteration is required for the definition of *measure*. In this case, we need to derive the relation $na=b$ from the relation $a+b=c$ where a, b, c are any three vectors and n is an arbitrary natural number. Let the first relation be designated by $M(abn)$, the second by $s(abc)$. The former is derived recursively from the latter as follows:

let $M(ab1)$ mean that a=b (i.e., $J(ab)$);

let $M(ab(n+1))$ mean that there is a vector x such that $M(axn) \cdot s(axb)$.

A general definition of this principle of iteration must be delayed until §7.

The sequence of natural numbers is crucial to Cantor's concept of *denumerability* which, as is well known, has given rise to *Richard's antinomy*. The standard version of this antinomy is as follows: The possible combinations of finitely many letters form a denumerable set. So, since each definite real number can be defined by finitely many words, there are only denumerably many real numbers—contradicting Cantor's classical proof that the reals are non-denumerable.[22] In discussing this antinomy, we shall substitute the concept "set of natural numbers" for "real number." As our underlying sphere of operation we take the natural numbers along with the single primitive relation S. Within this sphere, the natural numbers are, without exception, individuals. So we can do entirely without Pr. 5 of §2 when we construct the derived properties and relations. We shall, however, adopt the (not as yet completely defined) principle of iteration. It is then certain that the "production process" of the judgment schemes of the derived properties and relations can be arranged in such a way that these properties and relations are ordered in an "enumerated" sequence. According to §4, the one-dimensional sets of numbers correspond to the properties which appear in this sequence and, therefore, by means of the indicated process, all possible sets of natural numbers are also ordered in the same sort of denumerable sequence. This, it seems to me, is the real core of Richard's antinomy—which we have been able to uncover thanks to our concrete specification (supplied by the production principles) of the concept "finite definition." On the other hand, the denumerability of all sets of numbers is in fact refuted by Cantor's proof in an entirely differ-ence sense which, as I see it, applies only to mathematics. In our sphere of operation, there is no binary numerical relation $R\,(xy)$ which satisfies the following condition: for every (one-dimensional) set of numbers (corresponding to a derived prop-erty) there is a number a such that the set of numbers which corresponds to the property $R(xa)$ (i.e., the set of all numbers x which stand to a in the relation $R(xa)$) is identical to our initial set. Cantor's proof of this proposition amounts to considering the set of numbers corresponding to the property $\overline{R}(xx)$: certainly no number a can be associated with it in the required way.

If we adopt the concept of denumerability suggested by this proof, then *naturally there is no reason at all to assume that every infinite set must contain a denumerable subset*—a consequence from which I certainly do not shrink.[23]

§6. ITERATION OF THE MATHEMATICAL PROCESS
THE *CIRCULUS VITIOSUS* OF ANALYSIS

We started from a sphere of operation, i.e, from one or more individual categories of object, the "basic categories," and from certain individual, immediately exhibited "primitive" properties and relations which apply to objects of these categories. Each of the individual "blanks" of a relation (whether primitive or derived) is affiliated with a definite category of object in such a way that only an object of this category can be meaningfully used to fill the relevant blank. The category "natural number" together with the primitive relation S associated with it we call the *absolute* sphere of operation. We assume that the underlying sphere of operation contains this absolute one (in an obvious sense). The derived properties and relations arise from the primitive ones; and to each primitive or derived subject-ordered relation there corresponds, by virtue of the mathematical process, a one- or multi-dimensional set. The category to which such a set belongs is determined by the number of blanks in the relation from which the set arises and by the categories of object with which each of these blanks, in the established order of succession, is affiliated. For the moment, we wish to refer more precisely to all these properties and relations, and the corresponding sets and functional connections, as those of the *1st level*.

That *a*, *b*, ... form a system of elements of a set *M* is a relation between the objects *a*, *b*, ... and the set *M* which we wish to label by the letter ϵ. So one blank of this relation ϵ is affiliated with a definite category of 1st level sets, while the others are affiliated with the same basic categories with which the blanks of the sets belonging to the former category (or, rather, the relations to which they correspond) are affiliated. To the basic categories we now add the various categories of one- and multidimensional sets. To the primitive properties and relations which are affiliated with the basic categories we add the relation ϵ, which connects the objects of those categories with the sets. In this way, an expanded sphere of operation arises to which we can once more apply the "mathematical process." Thus we arrive at (one- or multi-dimensional) *"sets of the 2nd level"* whose blanks, in general, are affiliated in part with basic categories, in part with categories of first level sets. The mathematical process can be iterated in this way not just once, but arbitrarily often.

We must not overlook the fact that new sets whose blanks are all affiliated with the *basic* categories can appear on the *2nd* level. In particular, this will occur when, in using the principles of §2 to construct a relation *R* belonging to the "expanded" sphere of operation, certain blanks affiliated with categories of 1st level sets—and, further, all such blanks—are filled by "*", i.e., "there is." The existence of this ("2nd level") relation *R* is therefore linked to there being a set, *i.e, a 1st level relation* which satisfies a certain judgment scheme. *R* is clearly a relation of an entirely different sort from those of the 1st level. If, failing to bear in mind the distinctions between levels, one chose to speak here of a relation whose existence is linked to there being a relation such that . . .—one would trap oneself in an endless circle, in absurdities and contradictions entirely analogous to Russell's well-known paradox involving the set of all sets which are not members of themselves. (I maintain, and will presently show in some detail, that our current version of analysis spins in such circles constantly.) In the construction of the relation R, the existence concept is applied to the (1st level) *relations* in the same way that it is applied to the *objects* of the basic categories. For Pr. 6 (filling of a blank with "there is") can be applied both to blanks which are affiliated with a basic category and to those which are affiliated with a category of 1st level sets. Obviously, this use of the existence concept is to be restricted to the objects of the basic categories and, accordingly, in the iteration of the mathematical process *the two principles of closure 5 and 6 are to be applied only to blanks which are affiliated with a basic category.*[24] Given this "narrower procedure" it is clear that the sets and functional connections which hold between objects of the basic categories are exhausted by those of the 1st level; thus, new sets and functional connections of this kind are no longer added on the 2nd and higher levels. So if we comply with the narrower procedure, we need no longer distinguish the various levels, since the level on which a set stands is already determined by the category to which it belongs. For example, a three-dimensional set whose frist two blanks are affiliated with basic categories, but whose last is affiliated with one-dimensional sets of objects of a certain basic category, will have to be ranked on the 2nd level.[25]

Let us now apply these considerations to the foundations of analysis. In order not to waste time on non-essentials, we shall take the rational numbers as given, rather than beginning *ab ova* with the natural numbers.[26] So our underlying sphere of operation consists of 1. the category "natural number" and the relation S relevant to it; and 2. the category "rational number,"[27] as well as the ternary relations

$$s\,(xyz) : x+y=z, \quad p\,(xyz) : x \cdot y = z$$

and the property $P(x)$, meaning "x is positive," all of whose blanks are affiliated with the rational numbers.

Following Dedekind, we wish to characterize a *real number* a by means of the set of those rationals which are smaller than a. Thus we define a real number as a (one-dimensional) set a of rational numbers with the following properties:

a) if r is an element of a, then so is every rational number r' such that $r-r'$ is positive; b) for every element r of a there is a rational number r^*, also belonging to a, such that r^*-r is positive; c) a is non-empty, but not every rational number is an element of a.

The fact that some r is an element of a will also be expressed by the words "r is smaller than a" or by the symbols "$r < a$."

But how is the concept "set" to be understood here? In order to arrive at analysis, it is certainly not sufficient to apply the mathematical process just *once*; for in analysis we study not merely real numbers, but also sets of and functional connections between real numbers. Should we now comply with the "narrower iteration procedure" or not? If we do *not*, we arrive at a "hierarchical" version of analysis in which there are 1st, 2nd, 3rd, . . . level real numbers and, likewise, functions of various levels such that, e.g., a function of the 2nd level has a meaning only for arguments of the 1st and 2nd levels. Of course, this version of analysis would change into the one familiar to us if, wherever sets and functional connections are mentioned (particularly in connection with the expression "there is"), we could suppress the clause "of the 1st (or 2nd, . . .) level"—that is, if we were to proceed as though the 2nd level properties (which can be defined only on the basis of the *totality* of 1st level properties) also belonged to the primitive domain of 1st level properties. But precisely because of this, all definitions and proofs would become circular. For example, let M be a bounded set of 1st level real numbers. In order to construct its *least upper bound*, we must form a set g of rational numbers to which a rational number r belongs if and only if *there is* a 1st level real number belonging to M which is greater than r. This set g has properties a), b), c) and is therefore a real number, *but one of the 2nd level*, since in its definition "there is" appears in conection with "a 1st level real number" (i.e., "a set of 1st level rational numbers" or "a primitive or derived 1st level property").

The *circulus vitiosus*, which is cloaked by the hazy nature of the usual concept of set and function, but which we reveal here, is surely not an easily dispatched formal defect in the construction of analysis. Knowledge of its fundamental significance is something which, at this particular moment, cannot be conveyed to the reader by a lot of words. But the more distinctly the logical fabric of analysis is brought to givenness and the more deeply and completely the glance of consciousness penetrates it, the clearer it becomes that, given the current approach to foundational matters, every cell (so to speak) of this mighty organism is permeated by the poison of contradiction and that a thorough revision is necessary to remedy the situation.

A "hierarchical" version of analysis is artificial and useless. It loses sight of its proper object, i.e., number (cf. note 24). Clearly, we must take the other path—that is, we must restrict the existence concept to the basic categories (here, the natural and rational numbers) and must not apply it in connection with the system of properties and relations (or the sets, real numbers, and so on, corresponding to them).[28] In other words, the only natural strategy is *to abide by the narrower iteration procedure*. Further, only this procedure guarantees too that all concepts and results, quantities and operations of such a "precision analysis" are to be grasped as idealizations of analogues in a mathematics of approximation operating with "round numbers." This is of crucial significance with regard to *applications*. So a proposition such as the one mentioned above, that every bounded set of real numbers has a least upper bound, must certainly be abandoned. But such sacrifices should keep the path ahead clear of confusion.[29]

We have not yet determined what particular account we are to give of the concept *function*. Let us examine functions $x(t)$ whose independent variable t ranges over some category of object k (e.g. the natural numbers) but whose value is always a real number. Let $R(xt)$ be a binary relation whose blank x is affiliated with one-dimensional sets of rational numbers and whose blank t is affiliated with the category k. If for every object t of this category—or every element t of a one-dimensional set of such objects—there is one and only one set x of rational numbers with properties a), b), c) such that the relation R holds, then this "real number" x is a function of t. This is one possible formulation of the concept of function. But the following seems more natural. A real number x is given as a set of rational numbers which are characterized by a property they have in common. x will depend upon t if an arbitrary object t of the category k is a component of this property, ie., if that *property* arises from a binary *relation S* (oo)—whose first blank is affiliated with the category of rational numbers and whose second is affiliated with k—by filling the second blank with t. We then say that the set x of rational numbers corresponding to the property $S(o\,t)$ *depends on* or is a *function of t*. In particular, it can happen that for every object t of the category k or even just for every element t of a set consisting of objects of this category, the corresponding set x has the properties a), b), c) of real numbers. [This is a second possible formulation of the concept "function."]

Given the first formulation, not even the proposition "the sum of two functions is itself a function" would be correct. For instance, if $R(xt)$ and $R'(xt)$ are both relations which underlie functions and $S(xyz)$ *designates the relation* $x + y = z$ for real numbers, then the sum of these functions arises from a relation which would be formed as follows:

$$R(xt)\cdot R'(yt)\cdot S(xyz) = RR'S(xyzt); \quad RR'S(**z\,t).$$

Thus in its construction "$*$", i.e., "there is", must be used to fill a blank which is not affiliated with the basic categories. But if we are to abide by the narrower procedure, as we think we must, this is inadmissible. On the other hand, if we adopt the second formulation of the concept "function," then, clearly, the sum of two functions is itself a function. A further point emerged in the preceding section: Cantor's proof that the continuum is not denumerable (i.e., that there is no function $T(n)$ which maps the natural numbers onto the sets of natural numbers) assumes that the concept "function" is taken in the second sense. *Cauchy's convergence principle* is also true in the second formulation—and this is of decisive importance for the construction of analysis (cf. Chapter 2).

All these considerations lead us to the following concept of function (and once we become aware of it, we also immediately grasp its significance). Let the blanks of a relation, e.g. $R(uv|xyz)$, be divided into two ordered groups, the *dependent* blanks uv and the *independent* ones xyz. If we fill each independent blank with some object (x, y, z respectively) of the appropriate category, then R gives rise to a relation $R(oo|xyz)$ which possesses just the (ordered) "dependent" blanks; corresponding to this new relation is a two dimensional set F (xyz) which *depends on x, y, z* or is a *function of* the "values" of the independent variables x, y, z. (But while this set, the "function value," depends on the objects used in filling the blanks, i.e., on the values of the independent variables, the *category* to which this set belongs does *not* depend on these objects: rather, it is the category of those two-dimensional sets whose blanks are affiliated with the categories of the *dependent* variables u and v.) According to this account, the relations

$$R(uv | xyz) \text{ and } \in(uv\mathrm{F}\,(xyz))$$

are equivalent.

Example: Let R be the negation of the relation $\in$, i.e., $\bar{\in}(xX)$, in which the blank x is related to a definite category of object, the blank X to the category of one-dimensional sets of such objects. If we take x as dependent, X as independent, then we obtain the function $F(X)$ whose value for each X is the complement $\bar{X}$ of X. Here we have one of the simplest functions whose dependent "argument" belongs to the same category as the function value.

§7. PRINCIPLES OF SUBSTITUTION AND ITERATION

The natural concept of function on which our gaze is now directed permits the definitive general formulation of the principle of iteration mentioned in §5. So at this point, let us resume the general description of our system, taking it for granted that some arbitrary sphere of operation is given. The principle of *substitution* must precede that of iteration.

7. Let $R(uv\,|\,xyz)$ and $T(xwU)$ be two relations. Let the blank U (in T) be affiliated with the category of those two-dimensional sets whose own blanks are affiliated with the categories of the blanks u, v in R. And let the blanks designated by x in R and T both be affiliated with the same category of object. If I regard the blanks u,v in the relation R as the dependent ones, then R gives rise to the function $F(xyz)$ whose value is always a set belonging to the category of the blank U.[30] The principle of substitution now allows us to form the relation $T(xwF(xyz))$ (with the four blanks $xyzw$).

In order to express ourselves more simply, we have presented an example instead of giving a truly general formulation of 7. But it is clear, without further ado, how this principle is to be generally applied. As a limiting case, we also permit R to contain *no* independent blanks *at all*, so that the blank U in T will be replaced by a definite set defined by a given subject-ordered relation R. Thus the principle of substitution plays the same role for blanks which are affiliated with categories of sets as Pr. 5 in §2 does for those which are affiliated with a basic category. However, 7 permits one to substitute not just a *definite* set, but even a *function*.

8. (*Principle of Iteration*) Let $R(xx' \mid X)$ be a relation whose blanks are divided into the two ordered groups of the dependent xx' and the independent X. Let the independent blank X be affiliated with the category of those two-dimensional sets whose own blanks are affiliated with the same categories of object as the dependent blanks xx' in R. And let the function arising from R be designated by $F(X)$. The value of $F(X)$ is always a set of the same category as the value of the argument X (cf. the example at the end of §6). Using the substitution principle, we can form

$$R_2(xx' \mid X) = R(xx' \mid F(X))$$

(from which the iterated function $F(F(X))$ arises). Let R_1 be the relation R. From R_2 we can, in turn, form

$$R_3(xx' \mid X) = R_2(xx' \mid F(X))$$

and so on, in such a way that, for every natural number n,

$$R_{n+1}(xx' \mid X) = R_n(xx' \mid F(X))$$

where R_1 is R. *We regard $R_1, R_2, R_3, \ldots$ as those relations which arise from a single one*

$$R(n; xx' \mid X)$$

when the blank n which is affiliated with the category "natural number" is filled in successively by 1, 2, 3 ... .

This principle exploits the characteristic feature of the natural numbers, whose sequence is the general scheme of a procedure consisting in the iteration (endlessly repeated performance) of an elementary process. But the principle must still be extended in three respects in order for it to obtain its most general form. *First*, in addition to the independent blank X, others can occur in R which are not affected by the iteration. *Second*, in place of X, several blanks can be subjected to iteration simultaneously. For example, let

$$R(xx' \,|\, XY), \ \mathrm{T}(y \,|\, XY)$$

be two relations whose blanks are divided in the indicated way into dependent and independent. Then there arise two functions $\mathrm{G}(XY)$ and $\mathrm{H}(XY)$ respectively. Let the blank X be affiliated with the category of two-dimensional sets to which the function value G belongs. And let the blank Y be affiliated with the category of one-dimensional sets to which the function value H belongs. The conditions are thereby given for the construction of iterated relations:

$$R(1;xx' \,|\, XY) = R(xx' \,|\, XY);$$

$$R(n+1;xx' \,|\, XY) = R(n;xx' \,|\, \mathrm{G}(XY)\mathrm{H}(XY)).$$

$$\mathrm{T}(1;y \,|\, XY) = \mathrm{T}(y \,|\, XY);$$

$$\mathrm{T}(n+1;y \,|\, XY) = \mathrm{T}(n;y \,|\, \mathrm{G}(XY)\mathrm{H}(XY)).$$

Third, and finally, the function to be substituted at the nth step can itself depend on n. For example, let $R(xx'|Xn)$ be a relation whose last blank n is related to the category "natural number," while the same assumptions as above apply to the remaining blanks. Let the ensuing function be designated by F(Xn). The iteration which leads to the construction of the relation R^* is described by the formulas

$$R^*(xx' \mid X1) = R(xx' \mid X1);$$

$$R^*(xx' \mid X(n+1)) = R^*(xx' \mid \mathrm{F}(X(n+1))n).$$

The principle of iteration, by far the most complex of all, is the specifically *mathematical* one. As an example of its application, we consider integer multiples of a vector (mentioned in §5). Lower-case sans-serif letters will signify blanks which are affiliated with the category "vector." Blanks affiliated with the category "two-dimensional vector set" will be designated by capital sans-serif letters. Let X_0 denote the two-dimensional vector set which corresponds to the relation $\mathscr{S}$ (xyx). Let us form

$$\in(\mathsf{xzX}) \cdot \mathscr{S}(\mathsf{xzy}) \mid_{\mathsf{z}=*} = R(\mathsf{xy} \mid \mathsf{X});^{31}$$

from which we obtain $R(n;\mathsf{xy} \mid \mathsf{X})$ by iteration. Then $R(n;\mathsf{xy} \mid \mathsf{X}_0)$ is the relation $\mathsf{y}=n\mathsf{x}$.

Another example: We want to show that the *cardinality* of a set consisting of elements of a definite basic category is a function of this set and we want to construct this function. Small italic letters will be affiliated with the basic category just mentioned, capital italic letters with one-dimensional sets of objects of that category, capital script letters with the category of one-dimensional sets of such sets. $\mathscr{U}$ will denote the "universal set" of the last category. (In each category of set there is an empty and a universal set.) In the relation

$$\in(\gamma X) \cdot \bar{J}(x\gamma)$$

(i.e., y is an element of X different from x) let us consider y the dependent blank. This relation gives rise to the function $F(xX)$ ("the set of all elements of X different from x"). Let us substitute this function for U in $\in(U\mathscr{T})$ (thus obtaining $\in(F(xX)\mathscr{T})$) and let us form

$$\in(F(xX)\mathscr{T}) \cdot \in(xX)\big|_{x=*} = \mathscr{d}(X\,|\,\mathscr{T})$$

("there is an element x of X such that the elements of the set X different from x themselves constitute a set which is an element of $\mathscr{T}$"). The relation is iterated: $\mathscr{d}(n;X|\mathscr{T})$. Then $\mathscr{d}(n;X\,|\,\mathscr{U}) = \mathscr{a}(nX)$ means that X consists of at least n elements ("it is possible to delete an element from X n times in succession"). The empty set in the category of one-dimensional sets of natural numbers we call the "cardinal number 0"; the universal set we call the "cardinal number ∞"; the set of natural numbers $\leq n$ we call the "cardinal number n" (it is the standard set of n elements to which every other is related in the act of counting). If, in the relation $\mathscr{a}(nX)$, we consider n the dependent blank and X the independent one, then this relation gives rise to the function $N(X)=$ the cardinality of the set X (i.e., the cardinal number of its elements). It is 0 only for the empty set, ∞ for all infinite sets.[32] Thus we see precisely how the role played by the numbers as "cardinal numbers" in the determination of cardinality can be traced back to their primitive role—namely, the presentation of iteration in abstract purity.[33]

As the remarks concerning logic in §3 indicate, it is natural that an expansion of the *forms of inference* accompany the extension of the table of our principles of definition. Thus, in particular, the principle of iteration carries with it the Bernoullian "inference from n to $n+1$" (or "inference by complete induction").

§8. DEFINITIVE FORMULATION OF THE FOUNDATIONS.
INTRODUCTION OF IDEAL ELEMENTS

Since here I must guide us from a traditional presentation to a new one, we can only gain an unobstructed view by first struggling through the undergrowth. So our path will not be exactly the straightest. We see now, given that the principles of substitution and iteration are to be added, that we can no longer adhere to the notion of a production of relations and corresponding sets in separate *levels* (whereby on the 1st level there appear all sets whose elements belong to the basic categories, on the 2nd all those sets whose blanks are affiliated partly with basic categories and partly with categories of 1st level sets, and so on). For, clearly, "recoils" to earlier levels can result from certain applications of the principle of substitution.[34] But, because the application of the existence principle 6 is here restricted to the basic categories, these "recoils" will not lead to circular (and, therefore, meaningless) definitions. If we imagine, as is appropriate for an intuitive understanding, that the relations and corresponding sets are "produced" genetically, then this production will not occur through a "hierarchical" iteration of the mathematical process discussed in §4; rather it will occur in merely parallel individual acts (so to speak). This production process yields the totality of relations which can be derived from the primitive ones involving the basic categories and from $\in$ by applying the various principles of definition. One then, in a purely formal way, lets a "set" belonging to the realm of objects correspond to each of these (subject-ordered) relations.[35] There is no need to distinguish between logically equivalent relations or between the sets corresponding to them. But, initially [i.e., while the production process is still going on], we are not [always] in a position to decide whether sets corresponding to non-equivalent relations are to be identified. Accordingly, when defining relations one must never make use of the relation of equality between *sets*.[36] Now if R is any relation formed in this way, whose blanks are all affiliated with basic categories, then the assertion that some definite objects of these categories satisfy the relation R is meaningful and is either true or false. Thus it is also true or false that *every* system of elements which satisfies the relation R satisfies a certain relation R' of the same sort—and vice versa. If this is true, then one must identify the sets corresponding to these two relations. *After* this identification of the "1st level" sets has been accomplished (and here I continue to express myself genetically), one can proceed to relations whose blanks are affiliated partly with basic categories and partly with categories of 1st level sets. And it now makes perfectly good sense, in the case of any two such relations, to ask whether all the systems of elements which satisfy the one also satisfy the other. At this point, the identification of the 2nd level sets takes place—and so forth. *The essential thing is that in defining the relations no use is made of the concepts of the equality and existence of sets;* thereby, but also only thereby, do we avoid the meaninglessness of circular definition.

Let us withdraw all our provisional remarks (i.e., the whole of §4-§7). For we are now going to present the definitive formulation of the principles which are to govern the formation of relations.

I. The Starting Point

1) The foundation consists of, first, one or more individual categories of object (the "basic categories") and, second, certain individual immediately exhibited properties of and relations between the objects of the basic categories (the "primitive relations"). (Each blank of a relation, or its judgment scheme, is affiliated with a definite category of object in such a way that the judgment scheme yields a meaningful proposition only when each blank is filled by an object of that category.) Each blank of

each primitive relation is affiliated with a basic category. To the primitive relations we add the *identity* relation $J(xy)$ whose two blanks x and y are always affiliated with the same basic category. (The restriction to basic categories is very essential in this case.)

2) A *set* belonging to the realm of objects corresponds to each subject-ordered relation (with one or more blanks). For example, if the objects a,b,c (in that order) satisfy the ternary subject-ordered relation R, then we say that a,b,c form a system of elements of the corresponding set P. The category to which this set belongs is determined by the categories with which the first, second, and third blanks, respectively, of R are affiliated. As a further basic relation we introduce ϵ, which holds, for example, between a,b,c and P if a,b,c form a system of elements of the set P.

II. The General Principles

These are Pr. 1 through 4 of §2. It should again be noted that, when applying 2, 3, or 4, any blanks which are "made to coincide" must, of course, be affiliated with *the same* category of object.

III. Principles of Closure

These are Pr. 5 and 6 of §2, with the *restriction* that any blanks which are filled by immediately exhibited objects or by "there is" must be affiliated with a basic category.

IV. Principles of Substitution and Iteration

Pr. 7 of §7 is always included. And we adopt the most general form of the principle of iteration (i.e., 8 – including the three extensions we introduced) if (as we now wish to assume) our sphere of operation contains the "absolute one" (defined at the beginning of §6).

V. Identification, Sets, Functions

We now consider the primitive (properties and) relations mentioned in I and all the relations which arise from them through application of the principles listed under II, III, and IV. When a short expression is desirable, I shall call these primitive and derived relations the "delimited" ones.[37] Consider two subject-ordered relations of this kind whose blanks are affiliated, in the given order of succession, with the same categories of object. If every system of elements which satisfies the one also satisfies the other, and vice versa, then the two corresponding ("delimited") sets are identical to one another; otherwise they are distinct.

To every delimited relation *R* whose blanks are divided into the two ordered groups of the "dependent" and the "independent," there corresponds a *function F*. If each independent blank is filled by some object of the appropriate category, then the set which corresponds to the relation that thereby arises from *R* is the *value* of the function F for the "system of arguments" used in filling the blanks. Two (differently defined) functions are identical to one another if and only if their values for every system of arguments are identical to one another.

The "mathematically expanded" sphere of operation is thereby established. Joining the objects of the basic categories are objects of new ideal categories, i.e., the *sets* and *functions*; joining the primitive properties and relations are the relation ϵ and the relation which holds between a function F (of two arguments, say), the objects *a,b* and the value of the function F for the system of arguments *a,b*. This expanded sphere of operation includes a complete system of definite self-existent objects in the sense of §1. If we make this system the object of our investigation, then, in order to acquire a complete knowledge of it, we must rule on the truth or falsity of each *pertinent* judgment concerning this system. The current meaning of "pertinent judgment" emerges from §2: they are those judgments (in the proper sense, i.e., without blanks) which arise, through *unrestricted* application of principles 1 through 6 of §2, from the above basic relations of the expanded sphere of operation. As in §2, these basic relations are to be augmented by the identity relation (whose blanks can now be affiliated with *any* category of object of the expanded sphere, both blanks with the same category of course). With regard to Pr. 5, we should note that the nature of sets and functions indicates the manner in which they can be "immediately exhibited"—namely, by fixing the relations to which they correspond; i.e., by constructing these relations from the foundations mentioned in I by applying the principles of II, III, and IV. Questions we can ask about the system of objects mentioned above are, for example, whether either of two given sets is a subset of the other, whether a given function of a real variable is continuous, and so on. However, no "set of all subsets of a given set"[38] or "set of all continuous functions of a real variable" exist *in our sphere of operation*: such a set is not "delimited." The domain of possibly "non-delimited" judgments and judgment schemes which arise through *unrestricted* application of the principles of §2 already exceeds the scope of the criterion (listed under V) for the equality of two differently defined sets or functions.

I believe that the system described above is a simple, reasonable, adequate, and *consistent* foundation for the construction of analysis—in contrast to the currently accepted foundation which, because of its vague concept of set and function and its manner of applying the concepts of existence and identity (particularly to the real numbers), finds itself caught in a *circulus vitiosus*. Our principles for the formation of derived relations can be formulated as *axioms concerning sets and functions*; and, in fact, mathematics will proceed in such a way that it draws the logical consequences of these axioms.

In conclusion, some words about the *introduction of ideal elements* in mathematics. Let us take as an example the *ideals* in the theory of algebraic number fields. These are defined as follows. Every system **s** of finitely many algebraic integers determines an ideal (**s**). The proposition *U*, i.e., "the algebraic number a is divisible by the ideal (**s**)," is to mean that a certain relation $R(a\,\mathbf{s})$, which we need not explain further here, holds between a and **s**. The ideals exhaust their significance in their capacity as divisors of numbers, i.e., in their appearance within assertions of the above form *U*. Accordingly, two ideals (**s**) and (**s**′) are to be considered identical if and only if every number divisible by (**s**) is also divisible by (**s**′) and vice versa. Therefore the ideal (**s**) corresponds in the following way to the property $R(o\mathbf{s})$ of an algebraic number's standing in the relation R to the system **s**: The same ideal corresponds to two properties of this sort if and only if, despite possibly different intrinsic meanings, they in fact have the same extension. But, as we suggested earlier, this is precisely the essence of the concept "*set*"−in explicit contrast to the usual view of a set as the consciously surveyed "gathering" of its elements. So, in Dedekind's manner, we can straightway characterize the ideal as the set $M(\mathbf{s})$ corresponding to the property $R(o\mathbf{s})$. Since the introduction of ideal elements in mathematics always follows this same pattern−especially when it occurs by means of the so-called "definition by abstraction"[39]−*the concept of set and function can encompass all such "new" constructs.* Of course, as our own example shows, set-theoretic terminology will often be replaced by more suggestive modes of expression.

Concluding Remarks

The concept of function has two historical roots. *First*, this concept was suggested by the "natural dependencies" which prevail in the material world−the dependencies which consist, on the one hand, in the fact that conditions and states of real things are variable over *time*, the paradigmatic independent variable, on the other hand, in the *causal* connections between action and consequence. The arithmetical-algebraic operations form a *second*, and entirely independent, source of the concept "function." For, in bygone eras, analysis regarded a *function* as an expression formed from the independent variables by finitely many applications of the four primary rules of arithmetic and a few elementary transcendental ones. Of course, these elementary operations have never been clearly and fully defined. And the historical development of mathematics has again and again pushed beyond boundaries which were drawn much too narrowly (even though those responsible for this development were not always entirely aware of what they were doing).

These two independent sources of the concept of function join together in the concept "*law of nature.*" For in a law of nature, a natural dependence is represented as a function constructed in a purely conceptual-arithmetical way. Galileo's laws of falling bodies are the first great example. The modern development of mathematics has revealed that the algebraic principles of construction of earlier versions of analysis are much too narrow either for a general and logically natural construction of analysis or for the role which the concept "function" has to play in the formulation of the laws which govern material events. General *logical* principles of construction must replace the earlier *algebraic* ones. Renouncing such a construction altogether, as modern analysis (judging by the wording of its definitions) seems to have done, would mean losing oneself entirely in the fog; and, at the same time, the general notion of natural law would evaporate into emptiness. (But, happily, here too what one says and what one does are two different things.)

I may or may not have managed to fully uncover the requisite general logical principles of construction—which are based, on the one hand, on the concepts "and," "or," "not," and "there is," on the other, on the specifically mathematical concepts of set, function, and natural number (or iteration). (In any case, assembling these principles is not a matter of convention, but of logical discernment.) The one entirely certain thing is that the negative part of my remarks, i.e., the critique of the previous foundations of analysis and, in particular, the indication of the circularity in them, are all sound. And one must follow my path in order to discover a way out.

With the help of a tradition bound up with that complex of notions which even today enjoys absolute primacy in mathematics and which is connected above all with the names Dedekind and Cantor, I have discovered, traversed, and here set forth my own way out of this circle. Only after having done so did I become acquainted with the ideas of Frege and Russell which point in exactly the same direction. Both in his pioneering little treatise (1884) and in the detailed work (1893), Frege stresses emphatically that by a "set" he means merely the scope (i.e., extension) of a concept and by a "correspondence" merely the scope or, as he says, the "value-range" of a relation. Russell's theory of logical types[40] corresponds to the formation of levels mentioned in §6 and is motivated by his "vicious-circle principle": "No totality can contain members defined in terms of itself." Of course, Poincaré's very uncertain remarks about impredicative definitions should also be noted here.[41] But Frege, Russell, and Poincaré all neglect to mention what I regard as the crucial point, namely, that the principles of definition must be used to give a precise account of the sphere of the properties and relations to which the sets and mappings correspond. Russell's definition of the natural numbers as equivalence classes (a technique which he borrows from Frege) and his "Axiom of Reducibility" indicate clearly that, in spite of our agreement on certain matters, Russell and I are separated by a veritable abyss. So it is only to be expected that he discusses neither the "narrower procedure" nor the concept of function introduced at the end of §6.

My investigations began with an examination of Zermelo's axioms for set theory,[42] which constitute an exact and complete formulation of the foundations of the Dedekind-Cantor theory. Zermelo's explanation of the concept "definite set-theoretic predicate," which he employs in the crucial "Subset"–Axiom III,[43] appeared unsatisfactory to me. And in my effort to fix this concept more precisely, I was led to the principles of definition of §2.[44] My attempt to formulate these principles as axioms of set formation and to express the requirement that sets be formed only by finitely many applications of the principles of construction embodied in the axioms—and, indeed, to do this *without presupposing the concept of the natural numbers*—drove me to a vast and ever more complicated formulation but, unfortunately, not to any satisfactory result. Only when I had achieved certain general philosophical insights (which, incidentally, required that I renounce conventionalism), did I realize that I was wrestling with a scholastic pseudo-problem. And I became firmly convinced (in agreement with Poincaré, whose philosophical position I share in so few other respects) that *the idea of iteration, i.e., of the sequence of the natural numbers, is an ultimate foundation of mathematical thought*—in spite of Dedekind's "theory of chains" which seeks to give a logical foundation for definition and inference by complete induction without employing our intuition of the natural numbers. For if it is true that the basic concepts of set theory can be grasped only through this "pure" intuition, it is unnecessary and deceptive to turn around then and offer a set-theoretic foundation for the concept "natural number." Moreover, I must find the theory of chains guilty of a *circulus vitiosus*.[45] If we are to use our principles to erect a mathematical theory, we need a foundation—i.e., a basic category and a fundamental relation. As I see it, mathematics owes its greatness precisely to the fact that in nearly all its theorems what is essentially *infinite* is given a finite resolution. But this "infinitude" of the mathematical problems springs from the very foundation of mathematics—namely, *the infinite sequence of the natural numbers and the concept of existence relevant to it.* "Fermat's last theorem," for example, is intrinsically meaningful and either true or false. But I cannot rule on its truth or falsity by employing a systematic procedure for sequentially inserting all numbers in both sides of Fermat's equation. Even though, viewed in this light, this task is infinite, it will be reduced to a finite one by the mathematical proof (which, of course, in this notorious case, still eludes us).

If, as I have advocated, we give a precise meaning to the concept "set," then the following assertion gains a substantial content: "To every point of a line (given an origin and a unit of length) there corresponds a (distance-measuring) real number (= a set of rational numbers with the properties a), b), c), mentioned in §6) and vice versa." This assertion establishes a noteworthy connection between something given in the intuition of space and something constructed in a logical conceptual way. But, clearly, this assertion far exceeds everything which intuition teaches, or can teach, us about the continuum. For it does not offer a morphological description of what presents itself in intuition (that being, first and foremost, a fluid whole rather than a set of discrete elements). Instead, it gives an exact construal of an immediately given reality which, by its nature, is inexact—a process which is fundamental to all exact knowledge of (physical) reality and through which alone mathematics acquires significance for natural science. This continuum problem will be dealt with in greater detail in Chapter II.

It has become more common of late to regard the distinction between mathematics and formal logic as problematic. But, in our view, it is obvious that, however closely related the two may be, mathematics stands apart from logic as a science whose distinctiveness is clearly marked.

Chapter 2

The Concept of Number
and The Continuum
(Foundations of the Infinitesimal Calculus)

§1. NATURAL NUMBERS AND CARDINALITIES

In the domain of natural numbers, the basic relation S gives rise to the fundamental operations of *addition* and *multiplication* in the following way.[1]

If we begin with m and pass n times in succession from one number to the next, the number thus produced from n is $m+n$. More precisely: If **X** is an arbitrary set of pairs of natural numbers (= a two-dimensional set of natural numbers), then let $\in {}^*(pm \mid \mathbf{X})$ signify that q and m, where q is the number immediately preceding p, form a pair belonging to **X**:

$$\in {}^*(pm \mid \mathbf{X}) = \in (qm\mathbf{X}) \cdot S(qp) \big|_{q=*}.$$

This relation is iterated, thus forming $\in {}^*(pm \mid \mathbf{X}n)$, and **X** is replaced by the two-dimensional set corresponding to the identity relation $x=y$ (where x,y range over natural numbers). This gives rise to the relation $s(pmn)$, which is precisely that expressed by the equation $p=m+n$. One shows (by complete induction) that for any two numbers m and n there is one and only one p which stands to them in the relation s. The definition of addition states that

$$m+1=m', \quad m+n'=(m+n)'$$

(where the accent expresses passage to the immediately succeeding number).

51

The *associative law*

$$(p+m)+n=p+(m+n)$$

follows by induction on *n*. It says that, in the sequence of natural numbers, if I proceed from *p* first *m* and then *n* steps further, I arrive at the number which I reach from *p* in *m+n* steps. The proof of the *commutative* law

$$m+n=n+m$$

requires two steps: it follows by induction on *m* that the law holds for *n*=1; the general form of the law then follows by induction on *n*.

A number which I eventually reach if, beginning with *m*, I proceed from each number to its immediate successor, is called *greater than m* (in symbols: $> m$). Intuition suggests that the three possibilities

(1) $$n > m, \ n=m, \ m > n$$

form an exhaustive disjunction and that, in the first case, *one and only one* number *s* exists such that *m+s=n*. We can prove this too by induction, employing merely the basic facts that every number has a unique immediate successor and that every number other than 1 has a unique immediate predecessor. To start with, we let $n > m$ mean that there is a number *s* such that *m+s=n*. Then we first prove that for every *m* and *s*, $m+s \neq m$ (the sequence of numbers does not turn back on itself; i.e., no number is greater than itself). This is true for *m*=1 since

$$1+s=s+1=s' \neq 1$$

(for s' has an immediate predecessor, but 1 does not). If the theorem is true for m, then it is true for m' as well. For if $m'+s=m'$, then it follows that

$$m'=s+m'=(s+m)';$$

from which it follows [by the uniqueness of the immediate predecessor] that

$$m=s+m=m+s.$$

Further, if n is a natural number, then there is no number x such that neither $x \geq n$ nor $n > x$. This holds for $n=1$ because every number x is ≥ 1. And if it is true for n, then it is true for n' as well. For if $n > x$ or $n=x$, then $n' > x$; but if $x > n$ and $x=n+s$, then either $s=1$ and, therefore, $x=n'$ or $s > 1$, i.e., $s=t+1$, and so

$$x = n + s = n + (1 + t) = n' + t > n'.$$

From the associative law for addition we infer that if $p > n$ and $n > m$, then $p > m$. And from this it follows that no more than one of the possibilities listed in (1) can hold for a given choice of m and n and that subtraction is a single-valued operation, i.e., that the inequality $m + s > m + s^*$ follows from $s > s^*$.

The meaning of *multiplication* essentially derives from the formulas

$$1 \cdot m = m, \quad n' \cdot m = (n \cdot m) + m.$$

Our principles of definition allow us to form the relation $p = n \cdot m$ from the relation s in a manner entirely analogous to that sketched in Chapter 1, §7, for forming the integer multiples of a vector. The *distributive law*

$$(n_1 + n)\cdot m = (n_1 \cdot m) + (n\cdot m)$$

is immediately confirmed by induction on n; and it, in turn, allows the *associative law*

$$(n\cdot p)\cdot q = n\cdot(p\cdot q)$$

to be confirmed in the same way. The proof of the *commutative law* is somewhat more complicated. It depends on the two facts

$$n\cdot 1 = n \text{ and } n\cdot(m + 1) = (n\cdot m) + n,$$

both of which are proved by induction on n. These facts imply that $n\cdot x$ has the same value as $x\cdot n$ for $x = 1$ and that both products change in the same way (namely, increasing by n) if we proceed from x to its immediate successor x'. So the two products coincide for all x. One final thing to note about multiplication is that $s\cdot n < s^*\cdot n$ follows from $s < s^*$.

We call a set of numbers (more precisely, a one-dimensional set of natural numbers) a *cut* of the number sequence if there are no two numbers m and n such that $m < n$ and n is, but m is not, an element of the set. The empty and universal sets of natural numbers are both cuts in this sense. *If A is a cut which is neither the empty nor the universal set, then there is a number n such that A coincides with the set of all numbers $\le n$.* Proof: 1 is an element of A. (For given any element m of A, either $m = 1$ or $m > 1$; and in the latter case, if 1 were not an element of A, then, contrary to assumption, A would not be a cut.) Further, there is a number n which is an element of A, but whose immediate successor n' is not. For if no such number existed, it would follow by induction that every number was an element of A. This n has the required properties: namely, that every number $\le n$ is an element of A, but no number $> n$ is. We see then that the concept cut coincides exactly with that of *cardinal number* introduced in §7 of Chapter 1. Let the "cardinal number n" be designated in what follows by $\bar{n}$.

If the cut A is a subset of the cut B, but A is not identical to B, then we say that A is *smaller than B and B is greater than A*, and for this purpose we employ the same symbols " $<$ ", " $>$ " as above. Given any two distinct cuts, one will be the smaller, the other the greater, of the two. The empty set 0 is smaller, the universal set ∞ greater, than every other cut. And if $m < n$ (where m,n are natural numbers), then the same holds for the corresponding cardinal numbers, i.e., $\bar{m} < \bar{n}$.

The numbers can (in any sphere of operation) be used to determine the cardinality of sets of objects of any basic category. Let us designate objects of the basic category under consideration by lower-case script letters, one-dimensional sets of such objects by upper-case script letters, and natural numbers, as before, by lower-case italic letters. The relation $a(n\,\mathcal{X})$ which means that $\mathcal{X}$ consists of at least n elements, was explained in Chapter 1, §7.

If $\mathcal{X}$ consists of at least n' elements, then it also consists of at least n elements.

This is true for $n = 1$. If it holds for n, it holds for n' too. For if $\mathcal{X}$ consists of at least $(n')' = n' + 1$ elements, then there is an element x of $\mathcal{X}$ such that the set of all elements of $\mathcal{X}$ distinct from x consists of at least n' and, thus, of at least n elements. But, by the definition of the relation a, that means that $\mathcal{X}$ consists of at least n' elements.

If $\mathcal{X}$ does not consist of at least m elements, then neither does it consist of at least m + n elements.

By the above result, this is true for $n = 1$. It follows by induction on n that it is true for all numbers. This proposition can be expresed in the positive form: "If $m < p$ and $\mathcal{X}$ consists of at least p elements, then $\mathcal{X}$ also consists of at least m elements."

Or: "For any given $\mathcal{X}$, a cut is formed by the natural numbers n such that $a(n\mathcal{X})$." This cut is precisely the *cardinality* of $\mathcal{X}$. If the cardinality is n, then the relation a $(n\mathcal{X})$ holds, but $a(n'\mathcal{X})$ does not.

The following proposition can be proved by induction. *If $\mathcal{X}$ is a subset of $\mathcal{E}$ and $\mathcal{X}$ consists of at least n elements, then $\mathcal{E}$ also consists of at least n elements.* This implies that the cardinality of a part is no larger than the cardinality of the whole. So a set which has an infinite subset is itself infinite.

From the definition of the relation a it follows immediately that if a new element is added to a set $\mathcal{X}$ which consists of at least n elements, then the expanded set $\mathcal{X}'$ consists of at least n' elements. In particular, this is the case if $\mathcal{X}$ consists of exactly n elements, i.e., if the cardinality of $\mathcal{X}$ is n. Not quite so obvious is the outcome of the opposite procedure:

R. If an arbitrary element x is removed from a set $\mathcal{X}'$ which consists of at least n' elements, then there remains a set $\mathcal{X}$ consisting of at least n elements.

The definition of a implies merely that *there is* an element x_0 such that the set $\mathcal{X}_0$ produced by dropping x_0 from $\mathcal{X}'$ consists of at least n elements. Nonetheless, R holds in all cases. We shall demonstrate this with the help of the following lemma concerning the *substitution of elements*: If a new object (of the relevant category) is substituted for one of the elements of a set $\mathcal{X}$ which consists of at least n elements (all else remaining unchanged), then the modified set $\mathcal{X}^*$ also consists of at least n elements. This proposition holds for $n = 1$. Let us assume that it is true for the natural number n. Let $\mathcal{E}$ be a set which consists of at least n' elements. Then there is an element e of $\mathcal{E}$ such that the set $\mathcal{X}$ of all elements of $\mathcal{E}$ distinct from e consists of at least n elements. In order to transform $\mathcal{E}$ into $\mathcal{E}^*$, we now substitute in $\mathcal{E}$ an object e_0^*, distinct from the remaining elements, for an element e_0. We have to distinguish two cases: either $e_0 = e$ or

$e_0 \neq e$. In the first case, $\mathscr{E}^*$ arises from $\mathscr{X}$ through the addition of a new element e_0^* and, accordingly, $\mathscr{E}^*$ consists of at least n' elements. In the second case, we transform $\mathscr{X}$ into a new set $\mathscr{X}^*$ by substituting e_0^* for the element e_0 of $\mathscr{X}$. So, by the inductive hypothesis, $\mathscr{X}^*$ consists of at least n elements. $\mathscr{E}^*$ includes all the elements of $\mathscr{X}^*$ as well as the element e not occurring in $\mathscr{X}^*$. So in this case too, $\mathscr{E}^*$ consists of at least n' elements.

R follows immediately from this lemma. For $\mathscr{X}$ is produced from $\mathscr{X}_0$ through the substitution of x_0 for x;[2] so $\mathscr{X}$ consists of at least n elements, just as $\mathscr{X}_0$ does. We can also show that if a new element is added to a set consisting of exactly n elements, the result is a set consisting of exactly n' elements. And if an element is removed from a set $\mathscr{X}'$ consisting of exactly n' elements, there remains a set $\mathscr{X}$ consisting of exactly n elements (so if an element is removed from an infinite set, an infinite set remains). That this last proposition holds no matter which element of $\mathscr{X}'$ is removed is clearly the basis of the well known process of *counting* and of the fact that this process gives rise to a result which is independent of the order of the counting. Further, we see that if an element of a set consisting of exactly n elements is replaced by an object distinct from the remaining elements, then the new set also consists of exactly n elements. In other words, the cardinality of a set is independent of the nature of its elements. Finally, we can prove by induction that, given two disjoint sets of cardinality $\overline{m}$ and $\overline{n}$, the cardinality of their union is $\overline{m+n}$. (This is confirmed for $n = 1$ by the preceding result concerning the addition of a single new element to a set.)

If the basic category we happen to be considering is that of the natural numbers, then the relation a allows us to count *sets of natural numbers*. And, in this case, it makes sense to maintain that the cut $\overline{n}$ of the sequence of natural numbers consists of exactly n elements. (Proof by induction using the above results.) By applying the concept "cardinal number" to sets of natural numbers, we can also establish that, e.g., the cardinal number $p(n)$ of the prime natural numbers up to n, which is $< \bar{n}$, is a *function* of n in our precise sense. The same holds for all other "number-theoretic functions."

In this section we have indicated how the elementary truths about numbers can, by copious use of induction, be derived from the two "axioms": "every number has a unique immediate successor" and "every number other than 1 has a unique immediate predecessor."

§2. FRACTIONS AND RATIONAL NUMBERS

Fractions appear as *multipliers* in daily life and wherever they are used to measure magnitudes which can be added together. In the case of, say, the vectors on a line, the repeated addition of a vector to itself gives rise to the operation of vector *replication* (cf. Chapter 1, §7). So, for each natural number m, let ma signify the "m-fold" of the vector a, which is itself a definite vector. This operation has a unique inverse, *subdivison*. That is, if a is a vector and n is a natural number, then there is one and only one vector x $=$a/n such that nx $=$ a. *m*a/n, *the* "m/n-fold" of a results from the combined application of replication and subdivision. The fraction sign m/n is a symbol for this composite operation, in the sense that two fractions are the same if both lead, by the indicated operations, to the same result for every vector a. Instead of the "operation" which produces the vector

$$(2) \qquad \qquad y = mx/n$$

from the arbitrary vector x, we prefer to speak of the subject-ordered "relation" between x and y expressed by the equation (2) or by

$$(3) \qquad \qquad ny = mx.$$

Fractions are correlated with relations of this form in such a way that the same fraction corresponds to two relations with the same extension. Accordingly, the fraction m/n is none other than the *set of pairs of vectors* corresponding to the relation (3). *Multiplying* fractions means successively executing the corresponding vector operations. The laws of multiplication follow from the fundamental fact that vector replication and subdivision commute with each other. That fractions may be *added* depends on the fact that the operation (on a) expressed by

$$(ma/n) + (m^*a/n^*)$$

can be represented by a single fraction, which will be designated by the sum $(m/n) + (m^*/n^*)$. We see then that the addition and multiplication of fractions are outgrowths of the concrete applications in which we employ fractions.

It is not expedient to introduce special fractions for each domain of magnitudes. Since the laws of fractions are independent of the nature of the domain of magnitudes under consideration, it is more practical to define the fractions purely arithmetically, in such a way that, in each domain of magnitudes, they are fit to symbolize each of the infinitely many possible composite processes of replication and subdivision. A simple way of accomplishing this is to apply the preceding considerations to the *system of natural numbers*, which obviously constitutes a domain of magnitudes in which addition is possible. The development of the theory will not be impaired by the fact that, in this domain, the relation (3) does not always admit of a solution with respect to y. So the following construction arises.

Let the relation $a \cdot b = c$ (where a, b, c are natural numbers) be designated by $\rho(abc)$. Let us form the relation

$$\rho(mxz) \cdot \rho(nyz)\,|_{z=*}$$

(i.e., $mx = ny$). If we replace m and n by two definite natural numbers, then to the binary relation which thus arises there corresponds a set of pairs of natural numbers. We call this set the fraction m/n. And m/n is also the symbol for a definite function (the "fraction function") of the two independent arguments m and n. From now on we shall use the first few letters of the lower-case script alphabet to designate fractions. If x,y form a pair of elements of the fraction (i.e., the two-dimensional set) a, then we say that y stands to x in the ratio a. If $a = m/n$, then, in particular, m stands to n in the ratio a. Using the laws of multiplication for the natural numbers, we can prove that "two" fractions m/n and m^*/n^* are identical with one another if and only if

$$mn^* = m^*n.$$

The relation $a \cdot b = c$ means that if x stands to y in the ratio a and y stands to z in the ratio b, then x stands to z in the ratio c. We can prove that, for any two fractions a,b, there is always one and only one fraction c such that this relation holds between them. It is called the product of a and b and is designated by $a \cdot b$. (If $a = m/n$ and $b = m*/n*$, then

$$a \cdot b = mm^*/nn^*.)$$

The relation $a + b = c$ means that if x stands to z in the ratio a and y stands to z in the ratio b, then $x + y$ stands to z in the ratio c. Here too the existence and uniqueness of such a c can be proved for any a and b. (And, if $a = m/n$ and $b = m^*/n^*$, then

$$a + b = (mn^*+m^*n)/nn^*.)$$

The basic laws of addition and multiplication follow easily from these definitions.

We can use the addition of fractions to explain their replication and subdivision, as in the case of vectors. In the domain of fractions, subdivision turns out to be an everywhere-defined single-valued operation. And we can show that

$$mb/n \text{ and } (m/n) \cdot b$$

always coincide.

If a, b are any two fractions and there is a fraction c (i.e., there are two natural m and n) such that $b + c = a$ (i.e., $b + (m/n) = a$), then we say that a is *greater than* b (in symbols $a > b$) and b is *less than* a ($b < a$). There can be only one such fraction c. And the possibilities

$$a > b, \ a = b, \ a < b$$

form an exhaustive disjunction.

A complete isomorphism holds between the natural numbers m and the corresponding fractions $m/1$ with denominator 1: the relations of sum, product and greater-less which hold between the natural numbers are reflected exactly in the homonymous relations between the corresponding fractions. Nonetheless, we may not identify these natural numbers and fractions with one another: what is not identical, we cannot "make" identical. Still, we should note that in the application of numbers to measurement, the natural number m and the fraction $m/1$ both represent the same process, namely, "m-plication."

We could now proceed further along the path we envisioned in Chapter 1, §6. Having reached the level of the fractions, we could demolish the stairway which brought us here and, on this high ground establish the foundation for a new and more extensive edifice by admitting the natural numbers *and the fractions* as basic categories from the start. Of course, the category "two-dimensional sets of natural numbers" would then encompass the second basic category. And this comprehension of a basic category by a derived one certainly cannot be eliminated by any trick. (What is not distinct, we cannot make distinct.) But we can nevertheless "ignore" it, since all pertinent questions can be decided without establishing the identity or distinctness of objects which belong to distinct categories. An intricate sort of double-talk would then result. But, except for an expanded vocabulary, this approach offers us no more than does the direct continuation of our construction begun on the basis of the single basic category "natural number." Whatever linguistic convenience might attend the doubling up of our teminology can be secured much more simply in the following way, without our forsaking the foundation of "pure number theory."

No matter what the context, expressions of the form "*there is* a fraction with such and such properties" can only mean that there are two natural numbers m and n such that the fraction $a = m/n$ has the property under consideration. Let M be a two-dimensional set[3] of natural numbers such that if m,n form a pair belonging to M and the fraction $m^*/n^* = m/n$, then m^*,n^* also form a pair belonging to M. Such an M is called a *domain of fractions*; and if m,n form a pair belonging to M, then we say that the fraction $a = m/n$ *belongs to the domain M*. (The domain of fractions to which a and only a belongs is identical to a). We employ an analogous nomenclature when, in addition to the two blanks affiliated with the category "natural number," the set M also contains blanks affiliated with some other categories. By a "binary domain" of fractions we mean, naturally enough, a quaternary set of natural numbers such that if m,n,p,q form a system of elements of that set and

$$m^*/n^* = m/n, \; p^*/q^* = p/q$$

then m^*,n^*,p^*,q^* also form a system of elements of that set. If m,n,p,q form a system of elements of the binary domain, then $a = m/n$ and $b = p/q$ (in that order) form a pair of fractions belonging to the binary domain.

In any domain of magnitudes which (like the domain of vectors on a line) contains the singular magnitude 0 satisfying

$$a + 0 = 0 + a = a$$

and which contains, for every magnitude a, the opposite one -a satisfying

$$a + (-a) = 0,$$

the processes of replication and subdivision are joined by the operation of *"reversal"* which transforms a into -a and by the *"nullification"* process which converts every magnitude into 0. If these processes and their combinations with replication and subdivision are to be represented by "numbers," the realm of fractions must be expanded into that of the *rational numbers* through the inclusion of zero and the negatives.[4] We can obtain the rational numbers from the fractions in a purely arithmetic manner entirely analogous to our acquisition of the fractions from the natural numbers—we just let addition play the role in this new construciton which multiplication played in the old.

If a,b are two fractions, then the four-dimensional set corresponding to the relation

$$a + (u/v) = b + (x/y)$$

(where x,y,u,v are natural numbers) is a binary domain of fractions. We call it the *rational number $a - b$*. And the pair of fractions c,d belongs to it if and only if

$$a + d = b + c.$$

(We then say that c and d "differ" by the rational number a–b.) In what follows, rational numbers are designated by l, m, n, ... We can easily prove that

$$a - b = a' - b'$$

if and only if $a + b' = a' + b$.

In particular, the four-dimensional set of natural numbers defined by $x/y = u/v$ is a rational number, which we shall call 0 (there not being any need to worry about confusion with the cardinal number 0). Then $a - a = 0$. If a is a fraction, then the binary domain of fractions to which a pair of fractions c, d belongs if and only if

$$a + d = c$$

is a rational number, which we shall call $+a$. Likewise, the binary domain of fractions defined by

$$a + c = d$$

is a rational number, which we shall call $-a$. The particular ordering of the fractions by the relations "greater" and "less" implies that to every rational number distinct from 0 there corresponds one and only one fraction a (its "absolute value") such that the rational number is either $+a$ or $-a$. This fact allows us to distinguish *positive* and *negative* rational numbers.

The equation $l + m = n$ (where l, m, n are rational numbers) means that if the fractions a, b differ by l and b, c differ by m, then a, c differ by n. For any two rational numbers l, m there is always one and only one n which stands to them in this relation. This form of addition satisfies the associative and commutative laws. And it admits a unique inverse operation, i.e., subtraction. Moreover, $l + 0 = l$ for all rational numbers l.

This explanation implies that the rational numbers form a domain of "add-able" magnitudes in which replication, subdivision and "reversal" are everywhere-defined single-valued operations. If m is a natural number, $a = m/n$ a fraction, and l a rational number, then the symbols

(4) $$m\,l \text{ and } (m\,l)/n = a\,l$$

mean that the pairs of fractions constituting the binary domain $m\,l$ (or, respectively, $a\,l$) are just those obtained from the totality of pairs of fractions $c,\ d$ belonging to l by forming

$$mc,\ md \text{ or, respectively, } a \cdot c,\ a \cdot d \ .$$

If $l,\ m,\ n$ are rational numbers, then $l \cdot m = n$ means that *either* $l = 0$, $n = 0$ *or* there is a fraction a such that

$$l = +a,\ \ n = a\,m$$

or there is a fraction a such that

$$l = -a,\ \ n = -(a\,m)$$

Thus the multiplication of rational numbers is explained on the basis of addition. The arithmetical laws follow from the fact that the three elementary operations of replication, subdivision, and reversal all commute with one another. In the realm of the rational numbers, the four primary operations of arithmetic, with the exception of division by 0, are everywhere-defined and single-valued.

We say that l is greater than m if $l - m$ is positive.

In every context, the expression "*there is* a rational number with such and such a property" means that there are four natural number m, n, p, q such that the rational number

$$(m/n) - (p/q)$$

possesses that property. By analogy with the terminology introduced in the case of fractions, we now take a *domain of rational numbers* to be a binary domain of fractions such that if the pair of fractions a, b belongs to that binary domain and

$$a - b = a' - b',$$

then the pair of fractions a', b' also belongs to it. A domain of rational numbers is therefore a four-dimensional set of natural numbers.

§3. REAL NUMBERS

In constructing fractions and rational numbers, all we need are sets which arise as values of two definite functions

$$m/n \text{ and } (m/n) - (p/q)$$

when natural numbers are taken as arguments. But if we are to grasp the concept "real number" in full logical determinateness, we must first explain the notion of "*all possible*" sets of a given category. This is why we introduced our principles of definition. Only when we reach the problem of the real numbers must we turn back to the foundation, i.e., to the principles of the logical combination of judgments. For the analysis of the real numbers, right down to its logical roots, has a character entirely different from the arithmetic of the rationals. On the basis of our principles of definition, we are going to develop the initial elements of a theory of real numbers and functions. We shall then examine the relevance of this theory to the theory of magnitudes and to our intuition of the continuum.

First of all, let us continue our construction of pure number theory. A domain of rational numbers which includes a rational ℓ only if it includes all rational numbers $< \ell$ is called (as in the realm of the natural numbers) a *cut*. This cut is *open* if there is no greatest among the rational numbers belonging to it. An open cut of rational numbers distinct from both the null and the universal domains is called a *real number*. So the real numbers are certain four-dimensional sets of natural numbers. We shall refer to the category of these sets as "category RN." And we shall indicate objects which belong to this category by using lower-case bold letters. Since "being a real number" is a delimited property of such an object $\mathbf{x}$, the "set of all real numbers" exists in our sphere of operation.

Let $\mathbf{f}(t)$ be a *function* (of the sort defined in Chapter 1) whose argument t ranges over an arbitrary category K of objects and whose value always belongs to the category RN. Further, let T be a one-dimensional set of objects of the category K and, for every element t of T, let the function value $\mathbf{f}(t)$ be a real number. Then $\mathbf{f}$ is a *real-valued* function "in" the set T. Note that our principles of definition imply that a function like $\mathbf{f}(t)$ is always defined for *all* objects of a definite category. Clearly, however, it is possible that the function value is always a four-dimensional set of natural numbers, but not always a real number. The values of the argument which do yield real numbers form a delimited set. If, in particular, the category K is that of the natural numbers and $\mathbf{f}(t)$ is real-valued for all natural numbers t, then this function is called a *sequence of real numbers* (or, more briefly, a sequence). If K is the category RN and the above-mentioned set T is one whose elements are all real numbers, then **f** is a real function of a real variable in *T*. This same terminology applies (*mutatis mutandis*) to functions with several arguments.

The *sum* of two real numbers **x** and **y** is a *function* of **x** and **y**. We define it as follows. (Here the blanks **x** and **y** are affiliated with the category RN; m_1, n_1, m_2, n_2 with the category "natural number".) Let the relation

$$\mathcal{S}(m_1, n_1, m_2, n_2 \mid \mathbf{xy})$$

mean that there is a system of elements p_1, q_1, p_2, q_2 of **x** and a system of elements r_1, s_1, r_2, s_2 of **y** such that

$$(m_1/n_1) - (m_2/n_2) = ((p_1/q_1) - (p_2/q_2)) + ((r_1/s_1) - (r_2/s_2)).$$

If we divide the blanks in $\mathcal{S}$ into dependent and independent ones in the way indicated by the vertical line, then we obtain a *function* **x** + **y** whose value is always a four-dimensional set of natural numbers; more precisely, always a domain of rational numbers. In particular, if the arguments **x** and **y** are real numbers, then this domain is an open cut distinct from the null and the universal domains and, therefore, is itself a real number. The commutative and associative laws hold. And this form of addition admits a unique inverse operation, i.e., subtraction.

a < **b** means that **a** is a subset of **b**, but is not identical to **b**. Again, the three possibilities

$$\mathbf{a} > \mathbf{b}, \quad \mathbf{a} = \mathbf{b}, \quad \mathbf{b} < \mathbf{a} \; (\mathbf{a} > \mathbf{b})$$

form an exhaustive disjunction in the realm of the real numbers (though not in the more comprehensive realm of the domains of rational numbers). Each expresses a delimited relation between **a** and **b**, since **a** and **b** are four-dimensional sets of elements of the basic category "natural number".

If l is a rational number, then the rational numbers $< l$ form an open cut which is neither the null nor the universal domain and, so, is a real number. This "rational real number", which is a function of l, will be designated by $*l$. Distinct real numbers correspond in this way to distinct rationals. For every real number **a**,

$$\mathbf{a} + *0 = \mathbf{a}.$$

A real number $> *0$ is called *positive*; a real number $< *0$ is called *negative*.

The concept of the open cut turns into that of the *open remainder* if $<$ is replaced by $>$ throughout the definition. To every open cut **a** there corresponds as "complement" an open remainder; this is the domain to which a rational number belongs if and only if it is greater than a rational number which does not belong to the cut **a**. Conversely, to every open remainder there corresponds as its complement an open cut. This relation of complementation is symmetric.

The real numbers form a realm of "add-able" magnitudes in which replication, subdivision and "reversal" are everywhere-defined single-valued operations. The cuts (real numbers) $m\mathbf{x}$ and $a\mathbf{x}$ are obtained from the open cut **x** by multiplying every rational number belonging to **x** by the natural number m or by the fraction a, respectively. If l is an arbitrary rational number, then $\mathbf{y} = l\,\mathbf{x}$ if *either* $l = 0$, $\mathbf{y} = *0$, *or* there is a fraction a such that $l = +a$, $\mathbf{y} = a\mathbf{x}$, or there is a fraction a such that $l = -a$, $\mathbf{y} = -(a\mathbf{x})$. The product $l\,\mathbf{x}$ is a real number uniquely determined by l and **x**. Finally, if **a** and **x** are real numbers and **x** is positive, then $\mathbf{a} \cdot \mathbf{x}$ is the domain to which the rational number m belongs if and only if there is a rational number l belonging to **a** such that m belongs to the domain $l\,\mathbf{x}$. If **x** is negative, then **a** must here be replaced by the open remainder complementary to **a**. Then $\mathbf{a} \cdot \mathbf{x}$

will be a real number whenever **a** and **x** are. If **x** =*0, then **a**·**x** = *0. The essential properties of multiplication can easily be inferred from these definitions. And we can see that the product of two real numbers, just like the sum, is a function of these numbers; for all our definitions can be constructed step by step from the principles of Chapter 1 (even though, for the sake of brevity and intelligibility, I have chosen not to proceed so pedantically). The difference and the quotient of two real numbers are likewise functions of their two arguments. (Of course, the quotient is itself a real number only if the denominator is not *0.) For example, we can define the function "quotient" as follows. This function corresponds to the relation

$$Q(m_1, n_1, m_2, n_2 | \mathbf{x}\mathbf{y})$$

which means that **x,y** and the rational number

$$\mathit{l} = (m_1/n_1) - (m_2/n_2)$$

are such that *either* **y** is positive and $\mathit{l}\mathbf{y} < \mathbf{x}$ *or* **y** is negative and $\mathit{l}\mathbf{y} > \mathbf{x}$.

It now follows from the principle of substitution (Chapter 1, §7) that *if* **f** *and* **g** *are two real-valued functions defined in a given set, then their sum, difference, product, and quotient are such functions; again, however, this holds for the quotient only under the restriction that* **g** *must not take on the value* *0 *anywhere in its domain.* Here we have the simplest examples of how our *logical* principles of construction lead, in a particular case, to those *algebraic* ones which the older version of analysis dimly associates with the concept function. Two other such principles, which are constantly employed, follow immediately from Pr. 2 and 7: 1) From a function of several arguments ranging over a single category, we obtain a new function by making these arguments "*coincide*"; thus the function **f***(t,t)* arises from **f***(s,t)*. 2) In, say, a function which takes on real values for all real arguments, we can replace the argument with another real-valued function.

Our definitions of the fractions, rationals and reals are, of course, arbitrary to a certain extent. Their real significance lies in the role they play in the measurement of magnitudes of various sorts, that is, in their utility in the abstract representation of certain relations holding between magnitudes. But it is absolutely necessary that, to this end, the concept of number first be defined in a purely conceptual-arithmetical way. *Every* such definition is correct which supplies structures capable of characterizing the aforementioned "relations" between magnitudes. Nonetheless, it could be maintained that the definitions selected by us are the simplest and most natural ones which lead to this goal. Later on, we shall present a more detailed discussion of the connections between our theory of numbers and the theory of magnitudes.

For what follows, we shall require the function $\mathbf{x}^n$ *of the real number* **x** and the natural number *n*. It can be obtained by recursion, thanks to the fact that $\mathbf{x}^{n+1}\mathbf{y}$ results from $\mathbf{x}^n\mathbf{y}$ by substitution of **xy** for **y**. Thus let the relation ρ $(\ell|\mathbf{xy})$ mean that **x** and **y** are real numbers and

$$\ell = (m_1/n_1) - (m_2/n_2)$$

is an element of the cut **x·y**. (Actually, ℓ stands for the four blanks m_1, n_1, m_2, n_2 affiliated with the category "natural number.") If the blanks are divided into dependent and independent as indicated, then this relation gives rise to the function **x·y**. We iterate the relation by again and again substituting this function for the blank **y**, thus obtaining

$$\rho \, (\ell|\mathbf{xy};n).$$

Our final step is to take the real number *1 as the initial value of **y**. The function **x**^{n} corresponds to the relation which thus emerges.

This is a convenient point to add a discussion of the concept *algebraic number*. As is well known, a real number **a** is called "algebraic of degree at most n" if there are n rational numbers l_1, l_2, ..., l_n such that

$$\mathbf{a}^n = l_1\mathbf{a}^{n-1} + l_2\mathbf{a}^{n-2} + \ldots + l_n.$$

Accordingly, to be "algebraic of degree at most 3" is without doubt a delimited property of real numbers; and the same holds for every other definite natural number. But, at first glance, it does not appear to be the case that the proposition "**a** is algebraic of degree at most n" is the judgment scheme of a delimited relation between **a** and n. And the status of the property "being algebraic" (without any specification of degree) is thus also suspect. Rather it appears as though, in order to resolve this problem by brute force, we must introduce relations with an "indefinite" number of blanks (a very unfortunate step from a logical point of view) and must extend the principles of definition, particularly that of iteration, to such relations in extremely complicated ways. But this is not the case at all. The concept of algebraic number will serve as a good example of how our principles of definition allow us to fare quite well even under such apparently unfavorable circumstances.

Just as we were able to explain the power **a**^{n} through iteration of the product function, so, analogously, we account for a polynomial in **a** of degree n with rational coefficients by applying iteration to the function

$$(\mathbf{a}\cdot\mathbf{b})\text{-}*l$$

formed from

$$\mathbf{a}, \mathbf{b}, \ell$$

where ℓ is any rational number. This function arises from the relation

$$\mathscr{G}(m \mid \ell; \mathbf{a}, \mathbf{b})$$

which means that the sum of the rational numbers m, ℓ is an element of the cut $\mathbf{a} \cdot \mathbf{b}$. (Once again, ℓ and m each stand for four blanks affiliated with the category "natural number.") In what follows, blanks designated by L will be affiliated with two-dimensional sets of objects of the category RN. We form

$$\in (\mathbf{a}, (\mathbf{a} \cdot \mathbf{b}) - {}^*\ell; L)\big|_{\ell = *} = \triangle(\mathbf{a}, \mathbf{b} | L)$$

which means that there is a rational number ℓ such that $\mathbf{a}$ and $(\mathbf{a} \cdot \mathbf{b}) - {}^*\ell$ form a pair belonging to L. By dividing the blanks of $\triangle$ into dependent and independent in the indicated way, we set the stage for an iteration which yields the relation $\triangle(\mathbf{a}, \mathbf{b} | L; n)$, meaning that there are n rational numbers $\ell_1, \ell_2, ..., \ell_n$ such that $\mathbf{a}$ and

$$(5) \qquad \mathbf{a}^n \mathbf{b} - (\ell_1 \mathbf{a}^{n-1} + \ell_2 \mathbf{a}^{n-2} + ... + \ell_n)$$

form a pair belonging to L. Here one must imagine the expression (5) written thus:

$$\mathbf{a} \cdot (...(\mathbf{a} \cdot (\mathbf{a} \cdot (\mathbf{a} \cdot \mathbf{b} - {}^*\ell_1) - {}^*\ell_2) - {}^*\ell_3)...) - {}^*\ell_n.$$

Now we need only replace $\mathbf{b}$ with the real number $*1$, and $L = L(\mathbf{a}, \mathbf{b})$ with the two-dimensional set L_0 corresponding to the relation

$$\mathbf{a} \text{ is a real number and } \mathbf{b} = {}^*0.$$

The ensuing relation

$$\triangle(\mathbf{a},*1 \,|\, L_0;n) = \triangle(\mathbf{a},n)$$

means that **a** is algebraic of degree at most n. $\triangle(\mathbf{a},*)$ is the judgment scheme of the property "**a** is algebraic."

We introduce the *complex numbers* in the standard way as *pairs* of real numbers. In general, we understand the formation of a pair as follows. If, say, A is a three-dimensional set and B a two-dimensional one (of any category, as long as neither A nor B is the empty set of its category), then there is a five-dimensional set $A \cdot B$, of which r, s, t, c, d form a system of elements if and only if r, s, t constitute a system of elements of A and c, d one of B. (The existence of $A \cdot B$ follows by Pr. 3, without identification of blanks.) We call $A \cdot B$ the pair formed from A and B. If we leave the sets A and B each undetermined within its category, then this pair is a function of A and B. If A and B are sets of objects of the basic categories, then, conversely, the "terms" A and B are functions of the pair $A \cdot B$. For if $\mathscr{G}$ is an arbitrary set of the category to which $A \cdot B$ belongs, then consider the relation $R(rst \,|\, \mathscr{G})$, which means that there are two objects c and d such that r, s, t, c, d form a system of elements of $\mathscr{G}$. The function of $\mathscr{G}$ which arises from this relation supplies the first term A of the pair $A \cdot B$ when this pair itself is substituted for $\mathscr{G}$. So, if A and B are sets of objects of the basic categories, the concepts "function of A and B" and "function of the pair $A \cdot B$" essentially coincide, thanks to the principle of substitution. According to this account of the notion "pair", the complex numbers are eight-dimensional sets of natural numbers or, more precisely, binary domains of rational numbers.

§4. SEQUENCES. CONVERGENCE PRINCIPLE

Let $\mathbf{f}(n)$ be a sequence of real numbers and let $R\,(\ell \,|\, n)$ be the relation between the rational number ℓ and the natural number *n* from which the function $\mathbf{f}(n)$ arises. So, for every *n*, $\mathbf{f}(n)$ is the domain of those rational numbers which stand to *n* in the relation *R*. (Once again, $\ell = (p_1/q_1) - (p_2/q_2)$ represents the four blanks $p_1, q_1; p_2, q_2$ affiliated with the category "natural number.") We construct the *limit inferior* of this sequence in the familiar way. It is a domain of rational numbers to which ℓ belongs if and only if for some $\ell' > \ell$ there is a natural number *n* such that for all $m > n$ the relation $R(\ell' \mid m)$ holds. This domain (call it **a**) is an open cut and, so, is either a real number or the empty domain (the customary sign for which, in this connection, is $-\infty$) or the universal domain ($+\infty$). We write

$$\lim_{n\,=\,\cdot\infty} \inf \mathbf{f}(n) = \mathbf{a}$$

If *R* also signifies the "domain" (i.e., the five-dimensional set of natural numbers) corresponding to the relation $R(\ell \mid n)$, then we see that this lim inf is a *function* of *R*.[5]

From the existence of the *lim inf* it follows that *Cauchy's convergence principle* holds. As is well known, a sequence of real numbers **f** is called *convergent* if, for every fraction a, there is a natural number *n* such that for every *p* and *q* which are $> n$, the rational number $-a$ belongs to the domain $\mathbf{f}(p)\text{-}\mathbf{f}(q)$, but $+a$ does not. Further, we say that the sequence *converges to the real number* **c** if, for every fraction a, there is a natural number *n* such that for every $p > n$, the rational number $-a$ belongs to the domain $\mathbf{f}(p)\text{-}$ **c**, but $+a$ does not. In all these definitions, the logical expressions "there is" and "all" or "every" appear only in connection with natural numbers. The convergence principle reads: *The sequence $\mathbf{f}(n)$ converges to some real number **c** if and only if this sequence is convergent.* When convergence occurs, this **c** is identical with the lim inf of the sequence and is then called simply the *limit*.

All this carries over *mutatis mutandis* to *function sequences,* i.e., to cases where the relation $R(\ell \mid n)$, *which defines the sequence, contains blanks in addition to those indicated. For example, if there is an additional blank* **x** affiliated with the category RN, then R gives rise to the function $\mathbf{f}(\mathbf{x}n)$; and

$$\lim_{n \,=\, \infty} \inf \mathbf{f}(\mathbf{x}n) = \mathbf{g}(\mathbf{x})$$

is itself a function of the real argument **x**. Here we have the analytic *principle of construction via passage to the limit.* Usually, it is customary to write the argument n as an index. But, obviously, one must keep in mind that the construction cannot be performed on an infinite series of functions

$$\mathbf{f}_1(\mathbf{x}), \mathbf{f}_2(\mathbf{x}), \mathbf{f}_3(\mathbf{x}), \dots$$

mysteriously gathered together from here and there. For the principle of construction applies only to functions $\mathbf{f}_n(\mathbf{x})$ of **x** *and* n, functions formed in accordance with the principles of definition made quite precise in Chapter 1.

In place of Cauchy's convergence principle, various other theses, which are allegedly equivalent to it, have been chosen as the starting point for analysis. I shall cite some of them:

1. Exactly one number belongs to the intersection of a sequence of nested intervals whose lengths pass below every positive number. (This is of use, e.g., in dealing with the decimal expansion of a number.)

2. Given a monotone increasing sequence of real numbers whose members remain below a definite bound, there is a number to which the sequence converges.

3. *Dedekind's cut principle*: If A and B are two sets of real numbers such that every number belonging to A is smaller than every one belonging to B and if, for every fraction a, there is a number $\mathbf{x}$ belonging to A and a number $\mathbf{y}$ belonging to B such that $+a$ does not belong to the domain $\mathbf{y}$-$\mathbf{x}$, then there is one and only one real number $\mathbf{c}$ such that no number belonging to A is larger and none belonging to B is smaller than $\mathbf{c}$.

4. A bounded set of real numbers has a unique least upper and a unique greatest lower bound.

5. Every bounded infinite set of real numbers has an accumulation point.

Propositions 1 and 2 hold true in our own firmly grounded version of analysis. By a "sequence of nested intervals" we mean two sequences $\mathbf{f}(n)$ and $\mathbf{g}(n)$ such that

$$\mathbf{f}(n) \;<\; \mathbf{f}(n'),\; \mathbf{g}(n) \;>\; \mathbf{g}(n'),\; \mathbf{f}(n) <\; \mathbf{g}(n)$$

(where n' is the immediate successor of n). However, the remaining assertions 3 to 5 fail to hold as stated; although they do hold if the sets of *real* numbers mentioned in them are replaced by domains of *rationals*.

We formulate the so-called *Heine-Borel Theorem* as follows:

6. Consider a sequence of intervals $\triangle_n$. Let every real number of the "unit interval" $*0 \leq \mathbf{x} \leq *1$ lie in the *interior* of one of the intervals of this sequence. Then there is a natural number n such that every one of these real numbers already lies in the interior of one of the finitely many intervals $\triangle_1, \triangle_2, ... \triangle_n$.

This proposition also turns out to be true in our system if the concept "sequence of intervals" is interpreted in an appropriate way. For then a delimited relation $R(\ell,n)$ between ℓ and n is expressed by the proposition, "The real number $*\ell$ corresponding to the rational number ℓ is either negative or lies in the interior of one of the intervals $\triangle_1,\triangle_2,...\triangle_n$, and the same also holds for every rational number $<\ell$." If 6 were false, then the set corresponding to the property $R(\ell,*)$ would be an open cut to which all negative rational numbers, but certainly not the rational number 1, would belong; so this set would be a real number in the unit interval. If one considers a member of the sequence of intervals $\triangle_n$ which includes this real number in its interior (and, by assumption, there is one), then a contradiction can be generated.

On the other hand, the Heine-Borel Theorem becomes *false* if in it the given *sequence* of intervals is replaced by an arbitrary set of intervals or if the argument n, which is indicated by the index n in $\triangle_n$, is replaced by an argument not affiliated with the basic category "natural number". In particular, the following may not be maintained: Given two functions $\mathbf{f(x)}$ and $\mathbf{g(x)}$ which take on real values in the unit interval and which satisfy the inequality

$$\mathbf{f(x)} < \mathbf{x} < \mathbf{g(x)}$$

for all values of their arguments, there are finitely many real numbers $\mathbf{a_1,a_2,...,a_n}$ in the unit interval such that, for every number $\mathbf{x}$ also in the unit interval, there is a number among the $\mathbf{a}_i$ such that

$$\mathbf{f(a}_i) < \mathbf{x} < \mathbf{g(a}_i).$$

The failure of some of the fundamental theses which were hitherto employed constantly in deductions within analysis entails that the currently accepted definitions and proofs must, in some cases, undergo an alteration, but, in others, must be abandoned altogether. In this regard, the failure of thesis 4 turns out to have a particularly profound impact. One consequence is that the inference scheme known as "Dirichlet's Principle" cannot be upheld even in the more moderate formulation which takes Weierstrass's critique into account and which no longer maintains the existence of a "minimum," but only of a "greatest lower bound." Anyone for whom the patterns of thought of contemporary analysis have become habitual must also constantly keep in mind that the presence of an infinite set of real numbers does not in itself guarantee the existence of a sequence $\mathbf{f}(n)$ consisting exclusively of numbers of this set.

By constructing partial sums, we can derive the theory of *infinite series* (or sums) from the theory of sequences. Let $\mathbf{f}(n)$ be a sequence of real numbers and let $U(\ell \mid \mathbf{b}, n)$ signify the relation "$\mathbf{b}$ is a real number and the rational ℓ belongs to the cut $\mathbf{f}(n) + \mathbf{b}$," i.e., the relation from which the function $\mathbf{f}(n) + \mathbf{b}$ arises. By applying the principle of iteration (in its third extension, described on p. 37) to this relation, we form $V(\ell \mid \mathbf{b}, n)$ where

$$V(\ell \mid \mathbf{b}, 1) = U(\ell \mid \mathbf{b}, 1); \quad V(\ell \mid \mathbf{b}, n') = V(\ell \mid \mathbf{f}(n') + \mathbf{b}, n).$$

The sequence of real numbers $\mathbf{s}(n)$ arising from $V(\ell \mid {}^*0, n)$ can then be linked to the original one by the recursion formulas

$$\mathbf{s}(1) = \mathbf{f}(1); \quad \mathbf{s}(n+1) = \mathbf{s}(n) + \mathbf{f}(n+1).$$

The connection between series and sequences carries over *mutatis mutandis* to series whose terms are functions of one or several variables. Notice, for example, that the power $\mathbf{x}^n$, as defined in the preceding section, is a function of **x** and *n*. It follows that the partial sums of the *power series*

$$\Sigma\, \mathbf{f}(n)\mathbf{x}^n$$

form a function sequence if **f**(*n*) is a sequence of real numbers. Hence the limit of this sequence, where it exists, is a real valued function of **x**. Corresponding observations may be made about infinite products.

The *elementary functions*, above all the exponential function, can be defined by using any of the infinite processes usually employed for this purpose. The logarithm can be defined as the inverse of the (continuous monotone) exponential function. (Concerning inversion, see the next section.)

§5. CONTINUOUS FUNCTIONS

Here we want to consider a function **f(x)** which takes on real values for the real arguments **x** belonging to the unit interval. Let this function arise from the relation $R(\ell\,|\,\mathbf{x})$. The equation

$$\mathbf{y} = \mathbf{f}(\mathbf{x})$$

expresses a delimited relation between **x** and **y**; for it means that all and only those rational numbers ℓ which stand to **x** in the relation $R(\ell|\mathbf{x})$ belong to the domain **y** (and here the concept "all" is indeed employed only in connection with "rational numbers"). Thus, for a given **y**, those numbers **x** of the unit interval such that **f(x)** = **y**, as well as those such that

$$\mathbf{f(x)} > \mathbf{y}\ (\text{or } \mathbf{f(x)} < \mathbf{y}),$$

form a set of numbers which is a function of **y**. On the other hand, the *value range* of the function **f** is generally not a delimited set of numbers. And if **f** is a bounded function, it generally has *no* definite least upper and greatest lower bounds.

Our particular concern in this section will be the *continuous* functions. By $|\mathbf{x}| \leq a$ we shall mean that **x** is a real number such that the rational number $+a$ corresponding to the fraction a does not belong to the domain **x**, but every rational number less than $-a$ does. The well-known definition of continuity reads[6]: **f(x)** is continuous at the number **a** (lying in the unit interval) if, for every fraction a, there is a fraction b such that

$$|\mathbf{f(x)}-\mathbf{f(a)}| \leq a$$

for all real numbers **x** of the unit interval which satisfy the inequality

$$|\mathbf{x}-\mathbf{a}| \leq b.$$

We can see that being continuous at a value **a** is *not* a delimited property of a function (and therefore is dependent on a precise demarcation of the scope of the concept "real number"). In the next section, we plan to give an account of the great significance of this fact for analysis, both pure and applied.

f(x) is *continuous on the unit interval* if it is continuous at every value **a** of that interval. It is *uniformly continuous* there if, for every fraction a, there is a fraction b such that

$$|\mathbf{f(x)}-\mathbf{f(y)}| \leq a$$

for all real numbers **x,y** of the unit interval such that $|\mathbf{x}-\mathbf{y}| \leq b$.

We wish to prove the following fundamental propositions concerning continuous functions:

A. *A continuous function assumes all intermediate values*; i.e., if **f** is a continuous function and

$$\mathbf{f(a)} < \mathbf{v} < \mathbf{f(b)}$$

then there is a real number **c** between **a** and **b** (i.e., $\mathbf{a} < \mathbf{c} < \mathbf{b}$) such that $\mathbf{f(c)} = \mathbf{v}$.

B. *A function which is continuous on the unit interval has a maximum and a minimum there*; i.e., there are two arguments **a** and **b** such that the inequality

$$\mathbf{f(b)} \leq \mathbf{f(x)} \leq \mathbf{f(a)}$$

holds throughout the unit interval.

C. *A function which is continuous on the unit interval is uniformly continuous there.*

The standard proofs of these propositions must be modified here in this respect: if **f** is the given real-valued function which is continuous on the unit interval, then we must initially consider only the values of **f(x)** for *rational* arguments. So (recalling that the real number *ℓ is a function of the rational ℓ) we form

$$\mathbf{f(^*\ell)} = \mathbf{f^*(\ell)}.$$

Actually, $\mathbf{f^*(\ell)}$ stands for a function of four arguments affiliated with the category "natural number."

Proof of A. If we assume that **f(*0)** is negative and **f(*1)** is positive, then we need only prove that $\mathbf{f(c)} = {^*0}$ for some number **c** in the unit interval. We form the domain of rational numbers to which l belongs if and only if there is a rational number $l' > l$ in the unit interval such that $f^*(l')$ is negative. This domain is a real number **c**. We then show, in the familiar way, that, since **f** is continuous at **c**, $f(c)$ can be neither negative nor positive and hence is $= *0$. (The idea of the proof is to construct the *greatest* point at which **f** takes on the value *0.)

Proof of B. The least upper bound **m** of $f^*(l)$ in the unit interval is that domain of rational numbers to which m belongs if there is a rational number l in the unit interval such that $m < f^*(l)$. **m** is either a real number or the universal domain $(+\infty)$. Given that $f(x)$ is continuous, it follows immediately that for *all* real values of the argument **x**, not just the rational ones, the inequality $f(x) \leq m$ holds. If **x** is any real number $> *0$ and $\leq *1$, then we can likewise form the least upper bound **m(x)** of $f^*(l)$ where l ranges over the non-negative rationals belonging to the domain **x**; this least upper bound is a function of **x**. We now distinguish two cases: *Either* the least upper bound $m(*l) = m$ for every positive rational number $l \leq 1$, in which case we take **a** to be the real number *0. *Or* the opposite occurs, in which case we form the domain **a** of rational numbers to which l belongs if there is a positive rational number $l' > l$ (and ≤ 1) such that $m(*l') < m$; this domain is a real number. In either case, it follows immediately from the continuity of $f(x)$ at $x = a$ that $f(a)$ cannot be smaller than **m**; so it must be the case that $f(a) = m$. This also shows that m cannot be the universal domain $+\infty$. (We have constructed the smallest value **a** at which **f** attains its maximum. In the same way, one constructs a number **b** at which **f** takes on its minimum.)

A and *B* can be combined thus: The value range of a continuous function on a closed interval is itself a closed interval.

For the *proof of C* it is expedient to assume that $f(x)$ is not constant, but is $= f(*0)$ for negative **x** and $= f(*1)$ for $x > *1$; we can do so without loss of generality. Let **x** be a real number, a a fraction. We form the least upper bound of

(6) $$|\mathbf{f}^*(l)-\mathbf{f}^*(m)|$$

where l and m are rational numbers such that l belongs to the domain $\mathbf{x}$ and $|l-m| < a$. This least upper bound is a real-valued function $\mathbf{d}(\mathbf{x},a)$ of $\mathbf{x}$ and a. Let $\mathbf{d}(a)$ be the least upper bound that results when we drop the requirement that l belong to $\mathbf{x}$. Then

$$\mathbf{d}(a) \geq \mathbf{d}(b) \ (> {}^*0) \text{ if } a > b; \text{ and } \mathbf{d}(\mathbf{x},a) \leq \mathbf{d}(a).$$

We have to show that

$$\lim_{a \,=\, 0} \mathbf{d}(a) = {}^*0.$$

To this end, we form the domain $\mathbf{x}(a)$ of rational numbers to which l belongs if there is a rational number $l' > l$ such that

$$\mathbf{d}({}^*l',a) < \mathbf{d}(a).$$

$\mathbf{x}(a)$ is a real-valued function of a. If $\mathbf{b}$, $\mathbf{b}'$ are any two real numbers on opposite sides of $\mathbf{x}(a)$, i.e.,

$$\mathbf{b} < \mathbf{x}(a) < \mathbf{b}',$$

then $\mathbf{d}(a)$ is the least upper bound of (6) for rationals l, m such that

$$\mathbf{b} \leq {}^*l < \mathbf{b}'$$

(i.e., l belongs to $\mathbf{b}'$, but not to $\mathbf{b}$) and

$$|m - l| < a.$$

Let

$$\liminf_{n = \infty} \mathbf{x}(1/n) = \mathbf{a}$$

and let g be any fraction. Then, thanks to the continuity of $\mathbf{f}(\mathbf{x})$ at $\mathbf{x} = \mathbf{a}$, there are real numbers $\mathbf{b}$ and $\mathbf{b}'$, on opposite sides of $\mathbf{a}$, and there is a positive real number $\mathbf{e}$ such that

(7) $$|\mathbf{f}(\mathbf{x}) - \mathbf{f}(\mathbf{a})| \leq (1/2)\, g$$

holds for all $\mathbf{x}$ of the interval

$$\mathbf{b}\text{-}\mathbf{e} < \mathbf{x} < \mathbf{b}' + \mathbf{e}.$$

From (7) it follows that

(8) $\quad |\mathbf{f}(\mathbf{x})\text{–}\mathbf{f}(\mathbf{y})| \leq g$, if $\mathbf{b} \leq \mathbf{x} \leq \mathbf{b}'$ and $|\mathbf{y}\text{-}\mathbf{x}| < \mathbf{e}$

Further, there is a natural number n such that the rational number $+1/n$ belongs to the domain $\mathbf{e}$ and $\mathbf{x}(1/n)$ lies between $\mathbf{b}$ and $\mathbf{b}'$. Since the inequality (8) holds for the indicated $\mathbf{x}$ and $\mathbf{y}$, g cannot be smaller than the least upper bound $\mathbf{d}(1/n)$ pertaining to this n. Hence

$$\mathbf{d}(a) \leq g \text{ whenever } a \leq 1/n.$$

Proposition A can be extended to continuous functions of several real arguments. The *fundamental theorem of algebra* also holds in our version of analysis.

A function $f(x)$ generally has no *inverse*. In fact, it can lack one even if, for every real number y which belongs to a given set T, there is one and only one number x such that

$$y = f(x).$$

But the existence of an inverse can be proved when $f(x)$ is a function which is *continuous* and *monotone* on the unit interval. For example, let $f(x)$ be monotone increasing, i.e., let

$$f(a) < f(b)$$

hold whenever a and b are two real numbers of the unit interval of which a is the smaller. If y is any real number, then we form the domain of rational numbers to which every negative l belongs as well as every $l < 1$ such that $f^*(l) < y$. This domain is a real number and, indeed, is the value of a definite function $g(y)$ for the argument y. If we restrict the variables x and y to the intervals

$$^*0 \leq x \leq {}^*1,\ f(^*0) \leq y \leq f(^*1),\ \text{respectively,}$$

then the functions f and g are inverses of one another:

$$f(g(y)) = y;\ g(f(x)) = x.$$

In the realm of *continuous* functions, *differentiation* and *integration* serve as function-generating processes just as they do in contemporary analysis: no change in the foundations is required. Of course, things are not so simple in the case of the more far-reaching integration- and measure-theories of Riemann, Darboux, Cantor, Jordan, Lebesgue and Carathéodory.

§6. THE INTUITIVE AND THE MATHEMATICAL CONTINUUM

So far, we have built up pure number theory from the natural numbers in much the same way as contemporary arithmetic and analysis do. With the help of our principles of definition, we pushed forward step by step without once pausing for a sidelong glance. Now we want to stop and to reflect on what we have accomplished.

As we saw, the continuity of a function is *not* a delimited property; i.e., in order to decide whether a function defined with the help of our principles is continuous or not we have to inspect not just the totality of natural numbers, but also the totality of *sets* (more precisely, four-dimensional sets of natural numbers) which arise from an arbitrarily complex joint application of those principles. If we regard the principles of definition as an "*open*" system, i.e., if we reserve the right to extend them when necessary by making additions, then in general the question of whether a given function is continuous must also remain *open* (though we may attempt to resolve any *delimited* question). For a function which, within our current system, is continuous can lose this property if our principles of definition are expanded and, accordingly, the real numbers "presently" available are joined by others in whose construction the newly added principles of definition play a role.[7]

Suppose the function "position of a point-mass" can be represented as a function of time. Then we can make certain abstract assertions about this function: for example, that it is continuous; or more simply that, for all real values of its argument belonging to a certain interval, the function itself only takes on values belonging to a certain region. Let us compare the latter proposition with the intuitive finding which it supposedly expresses in the mathematical representation of reality, no matter how "objectivized," "idealized," or "schematized" that expression may be. For example, *I see* this pencil lying before me on the table throughout a certain period of time. This observation entitles me to assert that during a certain period this pencil was on the table; and even if my right to do so is not absolute, it is nonetheless reasonable and well grounded. It is obviously absurd to suppose that this right can be undermined by an "expansion of our principles of definition"—as if new moments of time, overlooked by my intuition, could be added to this interval, moments in which the pencil was, perhaps, in the vicinity of Sirius or who knows where. If the temporal continuum can be represented by a variable which "ranges over" the real numbers, then it appears to be determined thereby how narrowly or widely we must understand the concept "real number" and the decision about this must not be entrusted to *logical* deliberations over principles of definition and the like.

In order better to understand the relation between an intuitively given continuum and the concept of number (the above example having revealed the discrepancy between the two), let us stick to *time* as the most fundamental continuum. And in order to remain entirely within the domain of the immediately given, let us adhere to *phenomenal* time (rather than to objective time), i.e., to that constant form of my experiences of consciousness by virtue of which they appear to me to flow by successively. (By "experiences" I mean what I experience, exactly as I experience it. I do not mean real psychical or even physical processes which occur in a definite psychic-somatic individual, belong to a real world and, perhaps, correspond to the direct experiences.) In order to have some hope of connecting phenomenal time with the world of mathematical concepts, let us grant the ideal possibility that a rigidly punctual "now" can be placed within this species of time and that time-points can be exhibited. Given any two distinct time-points, one is the *earlier*, the other the *later*. Two time-points *A,B* of which *A* is the earlier define a *time span AB*; into it falls every time-point which is later than *A*, but earlier than *B*. The experiential content which fills the time span *AB* could "in itself," without in any way being other than it is, fall within some other time; the time span which it would fill there is *equal to* the span *AB*. This account of the temporally "equal" may be very controversial; but even so, I wish not to linger over it. So let it be granted that for any two time spans, the assertion that they are equal to one another has a precise sense grounded in the intuition of time. Measurement is thereby made possible. And we are enabled to construct a *mathematical theory of time* on the foundation we have indicated: i.e., the basic category "time-point," the binary relation "*A* is earlier than *B*" *and the quaternary relation "AB* is equal to *A'B'*" (taking the natural numbers and their basic relation *S* as our guide). The discrepancy which we discussed above would disappear if:

1. The immediate expression of the intuitive finding that during a certain period I saw the pencil lying there were construed in such a way that the phrase "during a certain period" was replaced by "in every time-point which falls within a certain time span *OE*." (I admit that this no longer reproduces what is intuitively present, but one will have to let it pass, *if it is really legitimate to dissolve a period into time-points*.)

2. *If P is a time-point, then the domain of rational numbers to which / belongs if and only if there is a time-point L earlier than P such that*

$$OL = /\ OE$$

can be constructed arithmetically in pure number theory on the basis of our principles of definition and is thus a real number in our sense. Further, taking the time span *OE* as our unit, it is not merely the case that to every point *P* there thus corresponds a definite real number as its "abscissa", but also conversely to every real number there corresponds a definite time-point.

If the time-points with their relations of "earlier" and "equal" can really furnish the foundation of a pure theory of time, then the intuition of time must suffice to determine whether this correspondence between time-points and real numbers holds or not. If it does not hold, then we should attempt to expand or modify our principles of definition in such a way that the desired agreement comes about. But should it prove unattainable even in this way, then a pure-arithmetical version of analysis would have no real value and a theory of the continuum would have to be established independently, alongside number theory. Be that as it may, the question "Is 2 true?" remains to be answered. Or, at least, similar fundamental questions demand a reply (such as: "Are Dedekind's cut principle and Cauchy's convergence principle applicable to time points?"). In confronting these questions we cannot avoid the concept of set (or sequence), no matter how we twist and turn; *and the scope of this concept depends on the principles of definition*!

Now, I think that everything we are demanding here is obvious *nonsense*: to *these* questions, the intuition of time provides no answer—just as a man makes no reply to questions which clearly are addressed to him by mistake and, therefore, when addressed to him, are unintelligible. So the theoretical clarification of the essence of time's continuous flow is not forthcoming. The category of the natural numbers can supply the foundation of a mathematical discipline. But perhaps the continuum cannot, since it fails to satisfy the requirements mentioned in Chapter 1, §1: as basic a notion as that of the point in the continuum lacks the required support in intuition. It is to the credit of Bergson's philosophy to have pointed out forcefully this deep division between the world of mathematical concepts and the immediately experienced continuity of phenomenal time ("la durée").[8]

Why is it that what is given in consciousness presents itself not simply as a being (like the logical being of concepts), but as an enduring and changing being-now, so that I can say, "This *now* is, but *now* no longer is"? If, in reflection, we extricate ourselves from this stream and place ourselves opposite the constant Now which spans a changing experiential content, treating this Now as an object, then it becomes a *flow* for us, in which we can place points. To every point there corresponds a definite experiential whole; and if consciousness stands at a certain point, then it has the corresponding experiential whole; only this *is*. But what is the origin of the concrete duration of each experience? If we hold fast to the individual points in their isolation from one another,[9] then only one answer is possible: To be sure, I only have experiences belonging to a given time-point; but among these experiences is a more or less clear *memory* whose intentional object is the experience that I had in a past time-point. (We shall pass over the problem of how this memory is supposed to come to be accurate.) Thus if I have, say, a visual perception of brief duration, then, in a moment A, I have not only this perceptual experience, but simultaneously the memories "of" the perceptual experiences of all past moments which fall within this brief period. But not only this; for in this moment A, I remember not only the *perceptual* experience in the moment B which occurred a short time earlier, but the *entire* experience of this moment B, and this in turn contains in itself, in addition to the perception, the memories of the experiences I had in all earlier moments. The continuous observation would thus consist of infinitely many mutually related systems of infinitely many memories, one packed inside another, the earlier being the "contained" one. But, clearly, we experience none of this; and besides, such a system of point-like moments of experience fitted endlessly into one another in the form of a completely apprehended unity is absurd. The view of a flow consisting of points and, therefore, also dissolving into points turns out to be false. Precisely what eludes us is the nature of the continuity, the flowing from point to point; in other words, the secret of how the continually enduring present can continually slip away into the receding past.

Each one of us, at every moment, directly experiences the true character of this temporal continuity. But, because of the genuine primitiveness of phenomenal time, we cannot put our experiences into words. So we shall content ourselves with the following description. What I am conscious of is for me both a being-now and, in its essence, something which, with its temporal position, slips away. In this way there arises the persisting factual existent, something ever new which endures and changes in consciousness. What disappears can reappear; not, of course, as an experience which I have over again, but as content of an (accurate) memory, having become something past. In the objective picture which I form of the course of life, such a past thing is to be opposed to what presently is as something earlier. So we can gather the following concerning objectively presented time:

1. an individual point in it is non-independent, i.e., is pure nothingness when taken by itself, and exists only as a "point of transition" (which, of course, can in no way be understood mathematically);

2. it is due to the essence of time (and not to contingent imperfections in our medium) that a fixed time-point cannot be exhibited in any way, that always only an *approximate*, never an *exact* determination is possible.[10]

Corresponding remarks apply to every intuitively given continuum; in particular, to the continuum of spatial extension.

When our experience has turned into a real process in a real world and our phenomenal time has spread itself out over this world and assumed a cosmic dimension, we are not satisfied with replacing the continuum by the exact concept of the real number, in spite of the essential and undeniable inexactness arising from what is given. For, as always, there is more at work here than heavy-handed schematizing or cognitive economizing devised for fulfilling our practical tasks and objectives. Here we discover genuine reason which lays bare the "Logos" dwelling in reality (just as purely as is possible for this consciousness which cannot "leap over its own shadow"). But to discuss this further cannot be our business here. Certainly, the intuitive and the mathematical continuum do not coincide; a deep chasm is fixed between them. But there are rational motives which impel us across from one into the other in our effort to comprehend the world,[11] the same rational motives which push science from the experientially constituted reality in which we live as natural human beings over toward the "truly objective", exact, non-qualitative, physical world—from the chromatic qualities of visual things, e.g., to the oscillations of the ether or the corresponding mathematical descriptions of electro-magnetic fields. So one might say that our construction of analysis contains a *theory of the continuum* which must establish its own reasonableness (beyond its mere logical consistency) in the same way as a physical theory. Within this theory the concept "real number" is the abstract scheme of the continuum with its infinite embedding of possible parts; the concept "function" is the scheme of the dependence of "overlapping" continua (a particular instance of which is given, e.g., by a moving point; i.e., the overlapping of a temporal continuum by a linear spatial one). I cannot substantiate it in greater depth here, but, given that proposition A of the preceding section is satisfied by our real numbers and (continuous) functions, it will be clear without further ado how a very essential part of such a rational justification is at hand. Evidence for this is that those numbers and functions allow us to give an exact account of what "motion" means in the world of physical objectivity.

Exact time- or space-points are not the ultimate, underlying, atomic elements of the duration or extension given to us in experience. On the contrary, only reason, which thoroughly penetrates what is experientially given, is able to grasp those exact ideas. And only in the arithmetico-analytic concept of the real number belonging to the purely formal sphere do those ideas thoroughly crystallize into full definiteness. Let us restrict our discussion of space to the geometry of straight lines. Even if one still wishes to erect a theory of time and space as an independent mathematico-axiomatic science, one must still heed the following.

1. The exhibition of a single point is impossible. Further, points are not individuals and, hence, cannot be characterized by their properties. (Whereas the "continuum" of the real numbers consists of genuine individuals, that of the time- or space-points is homogeneous.) Therefore, points and sets of points can be defined only relative to (i.e., as functions of) a coordinate system, never absolutely. (The coordinate system is the unavoidable residue of the eradication of the ego in that geometrico-physical world which reason sifts from the given using "objectivity" as its standard—a final scanty token in this objective sphere that existence is only given and *can* only be given as the intentional content of the processes of consciousness of a pure, sense-giving ego.)

2. The continuity axiom must guarantee that, given a unit span OE, a real number corresponds to every point P as abscissa and vice versa. It is due entirely to this axiom that all pertinent judgments (in whose formation Pr. 5 of Chapter 1, §2 is excluded) have a clear sense in spite of the circumstance mentioned in 1.

3. The system which, for the moment, we shall call "hyperanalysis" arises if, starting from the level attained in §3 of this chapter, we lay a new foundation for pure number theory, a foundation in which we admit the real numbers as a new basic category alongside the naturals. (In §2 of this chapter, we considered a similar move: namely, including the rationals among the basic categories.) This new system certainly does not coincide with our version of analysis. On the contrary, in hyperanalysis there are, e.g., more sets of real numbers than in analysis. For hyperanalysis admits sets in whose definition "there is" appears in connection with "a real number." *Thus, hyperanalysis contains neither Cauchy's convergence principle nor, in general, our theorems about continuous functions.* (These propositions are satisfied only by the functions and sequences whose existence was already guaranteed by analysis.) Therefore we must again and again resist the perpetually renewed temptation of resorting to a higher level than the basic stratum of the natural numbers. *Only analysis, not hyperanalysis, furnishes a useful theory of the continuum* and of the possible dependencies between overlapping continua. But if we do adopt hyperanalysis, the situation is as follows. Because of the axiom indicated in 2, if we are given a definite underlying coordinate system *OE*, there is a universal correspondence not only between points and real numbers, but also between sets of points, sets of sets of points—in fact, between all sets of the space or time theory—and all sets of *hyperanalysis*; or, expressed still more precisely, this correspondence holds between the sets of hyperanalysis and the functions of *OE* in the space or time theory. So the indicated axiom can surely not be replaced by Cauchy's convergence principle (which, of course, does not hold in hyperanalysis) or by any similar formula hitherto standard in the axiomatic construction of geometry. (We are here ignoring Hilbert's axiom of completeness.) Further, given the intractability of

hyperanalysis, we see that it is futile to pursue time theory and geometry as independent axiomatic sciences. Elementary geometry (i.e., that part of geometry which can be developed without the continuity axiom) can probably be constructed synthetically— but real *continuity-geometry can only be dealt with analytically*, i.e., by constructing analysis within pure number theory and then applying its theorems to geometry via a transfer principle based on the introduction of a coordinate system; this is the only way to acquire exact and rational definitions of the concepts "curve," "surface," etc. The following assertion belongs to our theory of the continuum: a portion of space, the surface which forms the boundary of this portion of space, a portion of this surface or, in turn, its boundary line are all structures such that the totality of the points falling within them can be constructed arithmetically as three-dimensional sets of real numbers. This assertion is similar to the claim that a real number corresponds to every point on a straight line: the one is just as little confirmed or refuted by what is immediately given as the other. But this assertion does follow from the conception of exact space-points. So the geometric axioms just have the function of formulating the above-mentioned transfer principle using certain relations which are to be considered immediately given.

From our point of view, both contemporary analysis and its principles are left hanging in a nebulous limbo half-way between intuition and the world of formal concepts —even though, under the mask of its vague presentations of set and function, analysis is able to pass itself off as a science operating in the formal-conceptual sphere. Nevertheless, we must grant that the superstructure analysis has erected is largely unaffected by this critique of its foundation and can easily be released from the clinging remnants of clay, once the fog has been dispersed.

The reflections contained in this section are, of course, only a slightly illuminating surrogate for a genuine philosophy of the continuum. But since no really penetrating treatment of this topic is at hand and since our task is mathematical rather than epistemological, the matter can rest here.

§7. MAGNITUDES AND THEIR MEASURES

Let us dwell a moment further on the above-mentioned transfer principle which links space and time with numbers. To fix a time-point P relative to the "unit span" OE *in a conceptual way* is to construct a relation $\mathscr{L}$ (OEP) from the basic relations "earlier" and "later" using the principles of definition (not including Pr. 5), where $\mathscr{L}$ is such that to every two time-points O and E, of which O is the earlier, there corresponds one and only one point P which satisfies this relation. If the points $O'E'P'$ also satisfy $\mathscr{L}$, then we say that P' stands in the same *ratio to $O'E'$* as P does to OE: this seems to us to be the original sense of the concept "ratio." Further, please notice that there cannot be two relations $\mathscr{L}$ and $\mathscr{L}$ * with different extensions (to which different three-dimensional sets of points correspond) such that the same point P stands to the unit span OE in both the ratios $\mathscr{L}$ and $\mathscr{L}^*$. For if this occurred, then a set distinct from both the empty and the universal sets would be formed by those time spans OE to which there corresponds a point P such that the relations $\mathscr{L}$ (OEP) and $\mathscr{L}$ *(OEP) both hold. However, not only all time-points, but also all time spans are qualitatively identical to one another in such a way that this set (which corresponds to a derived binary relation between points belonging to our sphere of operation, in whose production Pr. 5 is never employed) does not exist. The continuity axiom says that all these relations $\mathscr{L}$, which we are calling "ratios," or rather their extensions, can be represented one-to-one by real numbers.

The above account of the concept "ratio" reveals the true significance of the numbers for the measurement of magnitudes.

In order to make this account somewhat more precise, let us consider time spans rather than time-points. The theory which we shall develop is, from a more general view-point, also the theory of an arbitrary linear positive *magnitude*. We fix our starting point as follows.

Category of object: time spans. Primitive relations: 1) a=b. This "equality" of time spans, which satisfies the well known axioms of the "equal" (i.e., every span is equal to itself; if a=b, then b=a; if a=b and b=c, then a=c) must not be confused with identity. 2) a+b=c. This relation continues to obtain if the three spans a,b,c are each replaced by an equal. Further, if this relation holds between a,b,c and also between a,b,c′, then c=c′. And this form of addition satisfies the commutative and associative laws. The most important derived relation is that expressed by the formulas a < b and b > a, which mean that there is a span d such that a+d=b. Since, given any two distinct time-points, one is the earlier, the other the later, it follows that one and only one of the three possibilities

$$a < b, a = b, a > b$$

holds, where a and b are any time spans.

On this foundation we shall now construct a mathematical discipline (taking the natural numbers as our guide). We shall employ the fundamental propositions of the first chapter, but not, of course, the principle of construction 5. An additional basic fact, then, is that our field of operation is *homogeneous*, i.e., that no one-dimensional set of spans other than the empty and universal sets exists. Accordingly, it is impossible to fix an individual span absolutely in a conceptual way, i.e., by a characteristic property. Rather, a span can be determined only relative to another, on the basis of a binary relation between spans $R(a,b)$. It is easy to see that each such relation which holds between a and b continues to obtain if these spans are each replaced by an equal one.[12] By "ratio" or "proportion" we understand a binary relation between spans $R(a,b)$ such that each span a stands in this relation to one and, in the sense of equality, only one span b. When we pass from the formal-logical to the objective standpoint, we do not distinguish proportions which have equal extensions; i.e., we replace each proportion by the two-dimensional set of spans corresponding to it. We call this set the *measure* of the proportion. That a,b form a system of elements of this measure $\mathscr{L}$, we express by the formula

$$b = a\mathscr{L}$$

or in the words, "b stands to a in the ratio $\mathscr{L}$." Then the following holds: if two spans stand to one another both in the ratio $\mathscr{L}$ and in the ratio $\mathscr{L}^*$, then $\mathscr{L}$ coincides with $\mathscr{L}^*$. For otherwise those spans a such that $a\mathscr{L} = a\mathscr{L}^*$ would form a one-dimensional set of spans distinct from the empty and universal sets. Measures can be *multiplied* and *added*. The following equations explain these operations:

$$(a\mathscr{L})\mathscr{M} = a(\mathscr{L} \cdot \mathscr{M}); \ (a\mathscr{L}) + (a\mathscr{M}) = a(\mathscr{L} + \mathscr{M}).$$

The natural concept of measure which we have just described has nothing intrinsically to do with the numbers of pure number theory. But we see now that those "pure" numbers, above all the naturals, are the indispensable conceptual means for establishing a measure. The relation of equality $a = b$ is a proportion, whose measure we shall call 1. More generally, the relation $b = na$, where n is an arbitrary natural number, arises from addition—as we have already indicated several times. This relation is a proportion, whose measure, which is determined by the natural number n, shall itself be designated by n. The inverse relation $a = nb$ or $b = a/n$ is a proportion too. To establish this we must show that for every span a there is 1) one and 2) in the sense of equality, only one b such that a=nb. If c is an arbitrary span, then the property which 1) attributes to a certainly belongs to nc. But, since a property which belongs to *one* span is, because of homogeneity, common to all spans, the 1st assertion follows. The 2nd part follows from the fact that if b $<$ b', then nb $<$ nb'. Given that both b=na and b=a/n are proportions, we see that the equation

$$b=(ma)/n$$

expresses a proportion determined by *m* and *n*. The measure corresponding to this proportion depends entirely on the fraction $m/n = a$; so this measure will itself be designated by *a*. The addition and multiplication of measures corresponding to fractions are entirely parallel to the addition and multiplication of these fractions themselves. The Archimedean Axiom (which cannot be reduced to simpler facts) holds: that is, if a and b are any two spans, then there is a natural number *n* such that na $>$ b.

We saw above that two spans indeed stand in *at most one* ratio to one another. But is it also true that they always stand in *some* ratio? The continuity axiom implies that this question is to be answered "Yes," as the following considerations show. If a and b are any two spans, then that domain of fractions to which g belongs if and only if g a $>$ b (and which is an "open remainder" in the realm of fractions) can also be defined within pure arithmetic; i.e., this domain of fractions appears in pure number theory and can be represented by a positive real number in an immediately obvious way. If, conversely, **x** is an open remainder of fractions in pure number theory which is neither the empty nor the universal domain, then the relation b=**x**a, which says that g a $>$ b holds for those and only those fractions g belonging to **x**, is a proportion, whose measure, which is determined by **x**, shall itself be called **x**. Thus, the measures "coincide" with the positive real numbers. Addition and multiplication in the two realms are entirely parallel.

We have developed a system of measures for *time spans*. But since all abstract measures are represented one-to-one by these particular measures, our general understanding of the concept "measure" would not in any way be enriched by our constructing an independent, abstract theory of measurement. This general understanding of measure merely plays the role of a stable regulative idea.

§8. CURVES AND SURFACES

As an example of how analytic concepts enable us to formulate geometric presentations in an exact way, we wish to conclude these investigations into the continuum by discussing the concepts "two-dimensional curve" and "spatial surface."

In plane geometry, we must distinguish two entirely different presentations which are usually both designated by the word curve. In order to keep them separate, I use the expressions "line" and "curve." Roughly stated, we are concerned with the distinction between the roadway system of a city or a street-car "line" on the one hand and, on the other hand, the route (= "curve") which a pedestrian traverses in the streets of this city (and which is *in statu nascendi* during the time of the stroll) or, respectively, the path which a moving street-car describes. "Lines" appear as, e.g., *boundaries* of parts of a region of the plane; a "*curve*" is the *path* of a moving point. By dissolving the plane into isolated points, we enable a line to be grasped as a set of such points which is constituted in a definite way—or, still more precisely, if, by means of the transfer principle of analytic geometry, we represent the points of the plane by pairs of real numbers and if we maintain our faith in the omnipotence of the Logos, then the line can be grasped as a binary set of real numbers which appears in pure number theory and which corresponds to a definite binary relation between real numbers (an "implicit equation"). The totality of points of the plane which a point moving in this plane "crosses" is a line in our sense. However, this line (which can be called the "trace" or "track" of the movement) must be not be identified with the path of the point. If the tracks on which a freight train is to run are given, the train can still traverse very different paths, in particular, paths of very different lengths [depending on the course it takes at the switching points and on whether it moves both backwards and forwards]. It is essential to a "curve" (in the second sense) that it be exhibited only in a movement—as an abstract (non-independent) moment of it. But in order for a *movement* to be given exactly, the locus of the moving points together with its dependence on time must be represented by two functions of a real argument constructed within pure arithmetic; the values of the argument correspond to the time-points, but the values of the functions correspond to the 1st and 2nd coordinates of the locus ("parametric representation"). We shall restrict our attention to this proper concept of curve, which is also adopted in infinitesimal geometry. The path itself is a one-dimensional continuum of "*path-points.*" Each path-point is in a definite place, i.e., coincides with a definite point of the plane, but is not itself this point of the plane. The path-points, as "stages" of the movement, stand in the relations "earlier" and "later" to one another, just as time-points do; in the movement, the continuum of path-points *spreads over* the continuum of time-points in a continuous monotone manner. This conception allows the "path" to be separated, in a way, from the movement which produces it. Our interpretation carries over to curves in three-dimensional space; but it becomes most important when we define the concept "surface"—and in this difficult case, we shall carry out the definition in an entirely mathematical way.

We are interested in that concept of surface which is analogous to "curve" rather than "line," i.e., in the sort of surface which infinitesimal geometry seeks to reproduce through parametric representation. I maintain that one's definition of this concept of surface will embrace every possible sort of self-intersection, interpenetration, and the like only if one takes the surface to consist of "surface-points," i.e., *sui generis* elements which form a continuum spread out in a dual way [just as a curve is spread out over both space and time]—a continuum which we shall call the "surface in itself." But this surface is embedded in space and, therefore, a definite space-point corresponds to each surface-point as the spatial position in which the latter is located. In the standard parametric representation

(9) $$x=x(u,v), \quad y=y(u,v), \quad z=z(u,v)$$

the three real numbers x, y, z serve as Cartesian coordinates which characterize the space-point, while the numbers u, v serve as "Gaussian" coordinates which characterize the surface-point, and the functions mathematically establish the correspondence we mentioned. Still, as is well known, the representation of surface-points by pairs of numbers does not adequately capture the global connectivity properties of all surfaces. When we give the mathematical formulation, we replace the "surface in itself" by a set S (appearing in pure number theory) of objects whose category one may freely choose; the elements of this set are the surface-points. (We shall not discuss the "transfer principle" which, on the basis of intrinsic relations existing between the surface-points, leads us from the latter to these objects of pure analysis, much as the concept of coordinates leads us from space-points to triples of real numbers.) But how are we to grasp the continuous connectedness of these points which unites them into the two-dimensional surface? Once we have torn the continuum apart into isolated points, it is difficult to give an abstract representation of the connectivity arising from the non-independence of the individual points. I shall here follow essentially the procedure which I employed in Weyl (1913).[13]

In the temporal continuum, an individual point exists only as a "transition point." We acknowledge and employ this fact within analysis when, having transformed the point, in spite of its transitoriness, into an independent individual, i.e., a real number **a**, we consider it relative to the infinite sequence of its neighborhoods which are defined by the inequalities

$$|\mathbf{x}\text{-}\mathbf{a}| \ < \ 1/n \ (n = 1,2,3,...)$$

and which shrink monotonically to **a**. We employ this substitute for continuous connectedness when, in particular, we give an exact definition of continuity (i.e., of continuous functions). Pre-critical analysis hoped to capture the non-independence of surface-points by means of the concept "infinitely close." But this notion, which could not be employed in a consistent manner, had to give way, in contemporary analysis, to the infinite sequence of ever more narrow neighborhoods.[14] Accordingly, we say that a "surface in itself" is given if (in pure number theory) a definite set **S** is given (by means of a property which characterizes its elements, the surface-points) and, in addition, a relation $N(P,Q;n)$. If two surface-points P,Q stand to one another in this relation, then we say that Q lies in the nth neighborhood of P. We require that this relation satisfy certain conditions:

1) P lies in every neighborhood of P. And the $(n+1)$th neighborhood of P is a part of the nth neighborhood of P.

Just as the set of all real numbers (which corresponds to the property $R(\mathbf{u})$, "$\mathbf{u}$ is a real number", with the blank $\mathbf{u}$) serves in pure number theory as a model of the one-dimensional continuum, so the analogous binary set (which corresponds to the binary relation $R(\mathbf{u}) \cdot R(\mathbf{v})$ with the two blanks $\mathbf{u}$ and $\mathbf{v}$), the so-called "number plane," is the model of the two-dimensional manifold. Accordingly, we require that every neighborhood be capable of transformation, by continuous mapping, into the interior of the unit square

$$|\mathbf{u}| < 1, |\mathbf{v}| < 1$$

of this number plane; what "continuous mapping" means here can be established with the help of the concept "neighborhood." So the following is a precise formulation of this requirement:

2) If P_0 is a surface-point, then there are continuous functions

$$P = P(\mathbf{uv}), \mathbf{u} = \mathbf{u}(P), \mathbf{v} = \mathbf{v}(P)$$

($P(00) = P_0$) which are inverse to each other and which supply a one-to-one mapping of the 1st neighborhood N of P_0 onto the interior of the unit square K of the number plane. The requirement that the functions be continuous means, in the case of $P(\mathbf{uv})$, that if n is any natural number and $\mathbf{u},\mathbf{v}$ is a point of the unit square, then there is a fraction a such that, for all real numbers $\mathbf{u}',\mathbf{v}'$ which satisfy the conditions

$$|\mathbf{u}'\text{-}\mathbf{u}| < a, |\mathbf{v}'\text{-}\mathbf{v}| < a,$$

$P(\mathbf{u}'\mathbf{v}')$ lies in the nth neighborhood of $P(\mathbf{uv})$. In the case of the function $\mathbf{u}(P)$, it means that, for every point P of N and every fraction a, there is a natural number n such that every point P' which belongs to the nth neighborhood of P lies in N and satisfies the condition

$$| \mathbf{u}(P')\text{-}\mathbf{u}(P) | < a.$$

The meaning for $\mathbf{v}(P)$ is just like that for $\mathbf{u}(P)$.

If we are to establish propositions concerning functions which are continuous on our surface and are to do so in a manner similar to that employed in §5 (of this chapter) for functions of a real argument, then we must postulate that:

3) There is a function $P(\ell)$ of one or several arguments which are indicated by the symbol ℓ and are affiliated with the category "natural number." Further, $P(\ell)$ is a surface-point for every argument; and, for every surface-point P and natural number n, there is a ℓ such that $P(\ell)$ lies in the nth neighborhood of P (i.e., the values of $P(\ell)$ are "everywhere dense" on the surface).

Let me stress explicitly that **S** need not be a one-dimensional set; it can also be a multi-dimensional one. In the latter case, the symbol P represents several blanks; however, everything we have done here retains its significance entirely.

From the "surface in itself," we pass to the spatial surface. **S** is embedded in space, i.e., three functions (defined in pure arithmetic)

$$\mathbf{x} = \mathbf{x}(P), \; \mathbf{y} = \mathbf{y}(P), \; \mathbf{z} = \mathbf{z}(P)$$

—which are real-valued for all P and are continuous—assign each surface-point P its spatial position. (If one restricts oneself to a neighborhood of a point, that is, if one investigates the surface only "locally," then a continuous parametric representation of the surface of the standard form (9) arises immediately with the help of condition 2) given above.)

The reduction of continuous connectedness to the concept of neighborhood has a defect: when a relation $N(P,Q;n)$ establishes the *n*th neighborhood of *P*, much more occurs than is given by the continuous connectedness itself. In the case of the plane, we could, e.g., choose the interior of the circle of radius $1/n$ about a point as the *n*th neighborhood of that point, but the circle of radius $1/2^n$ would serve just as well; further, we could employ neighborhoods which are elliptical, square or some other shape in place of the circular ones. We put up with this arbitrariness (just as we do the arbitrariness of the objects which we choose as representatives of surface-points) because it is obvious that no clear-cut answer is yet at hand to the question of how we shall establish the link between the given and the mathematical in a perspicuous manner. (Again and again, the inescapable discrepancy between the genuine continuum and a set of isolated elements manifests itself.)

By way of supplement, let us close by giving the conditions under which two analytic spatial surfaces coincide, i.e., are representatives of the same spatial surface in the intuitive sense. Suppose we are given a surface **S** which is embedded in space by means of the functions

$$\mathbf{x}(P), \ \mathbf{y}(P), \ \mathbf{z}(P)$$

and suppose we are also given a second surface **S*** whose points are assigned their spatial positions by the functions

$$\mathbf{x}^*(P), \ \mathbf{y}^*(P), \ \mathbf{z}^*(P).$$

Further, suppose we possess two continuous[15] functions

$$(10) \qquad\qquad P^* = F^*(P), \ P = F(P^*)$$

which are inverse to each other and which establish a one-to-one mapping of the two sets **S** and **S*** onto one another such that, for any two points P, P^* linked by (10),

$$\mathbf{x}(P) = \mathbf{x}^*(P^*), \ \mathbf{y}(P) = \mathbf{y}^*(P^*), \ \mathbf{z}(P) = \mathbf{z}^*(P^*).$$

Then we say that the two spatial surfaces coincide with one another as a result of the transformation (10).

*　　　*　　　*

Here we break off the development of our subject. We have seen that a version of analysis can certainly be constructed from our principles and we have carried out the initial stages of this construction—as far as we thought necessary to grasp the Pythagorean problem fully. To the criticism that the intuition of the continuum in no way contains those logical principles on which we must rely for the exact definition of the concept "real number," we respond that the conceptual world of mathematics is so foreign to what the intuitive continuum presents to us that the demand for coincidence between the two must be dismissed as absurd. Nevertheless, those abstract schemata supplied us by mathematics must underlie the exact science of domains of objects in which continua play a role.

Appendix

The *circulus vitiosus* in the Current Foundation of Analysis[1]

(From a letter to O. Hölder)

I hope the following lines will satisfy your wish that I might set before you in the most direct way possible the *circulus vitiosus* which, in my treatise *The Continuum*, I accuse analysis of harboring.

The *sense* of a clearly and unambiguously defined concept may indeed always assign the appropriate *sphere of existence* to the objects which share the essence expressed in the concept. But this certainly does not imply that this concept is *extensionally determinate*, i.e., that it is meaningful to speak of the *existent* objects falling under it as an ideally closed aggregate which is intrinsically determined and demarcated. Suppose P is a property pertinent to the objects falling under a concept C. And suppose P has a clear and unambiguous sense. Then the proposition "a has the property P" (where a is an arbitrary object which falls under C) affirms an entirely definite state of affairs which either does or does not obtain; this judgment is intrinsically true or untrue–steadfastly; a standpoint intermediate to these two opposites being impossible. But if the concept C is extensionally determinate, then not only the question "Does a have the property P?" (where a is an arbitrary object falling under C), but also the existential question "*Is there* an object falling under C which has the property P?", possesses a sense which is intrinsically clear. Corresponding remarks apply to relations. If C is extensionally determinate, it makes no difference whether the

109

sense of the property or relation at issue is exhibited immediately in intuition or whether, by means of logical operations, it is constructed out of properties and relations whose sense is intuitively given. (Of course, a thorough phenomenological analysis of the concept "*existence*" is needed here; but the preceding remarks may be sufficient for us to reach an understanding.)

The intuition of iteration assures us that *the concept "natural number" is extensionally determinate*. (Certainly, every version of arithmetic must extract *this* basic fact from intuition.) However, the universal concept "object" is not extensionally determinate—nor is the concept "property," nor even just "property of natural numbers." One can even prove the last assertion, if its obviousness is not immediately conceded. That is, let a definite sphere **k** of properties of natural numbers be marked out in such a way that the concept "**k**-property" is extensionally determinate. Then it is immediately possible to define properties of natural numbers which lie outside this sphere, as follows: If A signifies any property of properties of natural numbers, then consider the property P_A which belongs to a natural number x if and only if *there is* a **k**-property of the kind A which belongs to the number x—this property P_A most certainly differs in sense from every **k**-property. That is not to say that a **k**-property cannot be extensionally equivalent to a property like P_A. I call two properties (of natural numbers) *extensionally equivalent* if every number which has the one also possesses the other, and vice versa. A set corresponds to each property in such a way that the same set corresponds to extensionally equivalent properties. (This is the correct relation of the concepts "property" and "set". The failure to recognize that the *sense* of a concept is logically prior to its *extension* is widespread today; even the foundations of contemporary set theory are afflicted with this malady. It seems to spring from empiricism's peculiar theory of abstraction; for arguments against which, see the brief but striking remarks in

Fichte (1912, 6:133 ff.) and the more careful exposition in Husserl (1913a: 106-224). Of course, whoever wishes to formalize logic, but not to gain *insight* into it—and formalizing is indeed the disease to which a mathematician is most prey—will profit neither from Husserl nor, certainly, from Fichte.)

If we apply the preceding remarks to the concept "rational number" rather than to the concept "natural number" (in the justifiable certainty that this new concept too is extensionally determinate) and if we, along with Dedekind, conceive of a real number as a (specially constituted) set of rational numbers, then we realize that *the concept "real number" is not extensionally determinate*. Indeed, in Hölder (1892) you yourself state (in a footnote on p. 594) that you are in complete agreement with this assertion, a mere understanding of it being sufficient to preclude its rejection. But it is generally believed that this indeterminateness does not significantly affect the foundation of analysis, since the *sense* of the concept "real number" can be defined with adequate clarity: i.e., if any property of rational numbers (of a certain sort) is given, then a real number is also given thereby which separates the rational numbers sharing in this property from the rest. However, I would like to confirm once again that this view is completely wrong by analyzing the proposition that every bounded set of real numbers has a least upper bound.

A real number is a set of rationals which corresponds to a definite property of rational numbers. So a set of real numbers corresponds to a property A of properties of rational numbers. The least upper bound of such a set of real numbers is itself the set of those rational numbers x which possess a certain property P_A, i.e., the property that *there is a property of the kind A which* belongs to the number x. But such an explanation, which links the existence of a property P_A to there being (in general and without restriction) a property such that ..., is *clearly meaningless*;

for the concept "property of rational numbers" is not extensionally determinate. The explanation acquires a content only if the general concept "property" is narrowed to an extensionally determinate concept "**k**-property". Let this be done. And suppose that the concept "real number" is correspondingly restricted. When the explanation of P_A is modified in this way, P_A becomes a property whose *sense* most certainly places it outside the sphere of the **k**-properties. P_A may well be extensionally equivalent to a **k**-property and then, but also only then, a real number, the least upper bound, would correspond to P_A . Clearly, however, it is extraordinarily unlikely that it is possible, in an exact way, to set down an extensionally determinate concept "**k**-property" such that each property P_A , whose definition involves the *totality* of **k**-properties as indicated above, is extensionally equivalent to a **k**-property. In any case, *not even the shadow of a proof* of such a possibility exists; but precisely this proof would have to be effected in order for the assertion of the least upper bound's existence to *receive a sense in all cases and be universally true*.

If the usual explanations of such fundamental concepts of analysis as "least upper bound," "continuity," etc. thus lack a comprehensible sense as long as the general concept of property (and relation) is not restricted to an extensionally determinate one, "**k**-property", then the question arises of how such a restriction is to occur. The recent history of mathematics leaves no doubt concerning the answer: one is to restrict oneself to those properties and relations which can be *defined in a purely logical way* on the basis of the few properties and relations which are given immediately in intuition along with the relevant categories of object. (In the case of the natural numbers, we are given just the relation "is the immediate successor of".) I have attempted to give a precise formulation of the principles of this construction. It is probably not necessary to repeat explicitly that it would be meaningless to include among these principles an assertion such as the following: If A is a property of properties, then one may form that property P_A which belongs to an object x if and only if there is a property constructed by means of these principles which belongs to x and itself possesses the property A. That would be a blatant *circulus vitiosus*; yet our current version of analysis commits this error and I consider this ground for censure. It seems to me that my principles of definition constitute the major part of a *"pure syntax of relations"*[2] on which pure logic must rely when it has to work out the conditions under which two properties or relations formed by logical construction are equivalent. What sort of new relations will unfold before our intuition in the development of the life of the mind can certainly not be anticipated *a priori*. But I think it likely that the *principles of logical construction*, by means of which we derive composite relations from the primitive ones, can be set down once and for all (just like the elementary forms of logical inference). Such a venture does not impinge on the freedom of the mind. I do not presently insist that the completeness of my list be acknowledged, although, on the basis of logical reflection and relying on the vast constructive resources offered by the recent history of mathematics, I maintain that this completeness is virtually guaranteed.

The method of conceptual construction constitutes the essence of mathematical-physical cognition (as should have been clear since the time of Galileo and Descartes). So I hope that analysis too will find its way back to this method which it was about to abandon in favor of an entirely vague universality. The grounding of analysis is linked as closely as possible with applications, above all with physics. The *sense* of cognition directed toward the physical world thoroughly eludes me if I am not able to anchor the concepts "number," "set," and "function" in logical principles of construction in the manner I attempted in my treatise.

Allow me to make a few more remarks about these principles–remarks which should clarify the final standpoint to which I would like to lead the reader of *The Continuum* in the hope that he will then consider it self-evident that my foundational approach really does break the circle I have criticized. The principles of construction fall into two groups, the "logical" (Chapter 1, §2) and the specifically mathematical (Chapter 1, §7). I am going to focus on the latter. Mediating between these two groups means introducing the relation ∈ (in which case, a proposition like "the rose is red," which primitively predicates a property of the rose, is interpreted as asserting that the relation ∈ of "having" holds between the rose and the property red). In the early stages of my presentation, this transition also depended on the concepts of (one- and multidimensional) *set* and *function*; and the conception which drove us forward was that of the *iteration* of the construction process–that process being guided by the six "logical" principles. It seemed to me that this corresponds to the way ideas naturally develop, even though such a "dialectical" presentation, which again and again raises what has gone before to a higher standpoint, is not customary in mathematics. But in the systematic construction at which I finally arrive (Chapter 1, §8)–and this probably should have been emphasized even more strongly–*the conception of the iteration* [of the construction process] *is again dropped entirely* and the concept of set and function must be deferred much longer than was done originally (namely until the end: it finds its niche only under V on p.43).

Let us consider, say, the ternary relation ∈ (xy,Z) ("xy stand in the relation Z to one another") in which the blanks xy are affiliated with the same basic category, while Z is affiliated with the category of binary relations between objects of this basic category. And, in accordance with note 4 of Chapter 1, let us represent the scheme of this relation by a wooden plate with one

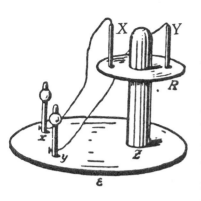

large and two small pegs corresponding to the blanks Z and xy respectively. The objects of the basic category are represented by balls, each equipped with a hole so that they can be stuck on the small pegs to represent filling the blanks xy. Suppose all this has been done. The blank Z in ϵ must be filled by a binary relation R. But this in turn is represented by a plate with two pegs, a plate which, moreover, must be equipped with a hole as big as the large peg in ϵ. If this plate is stuck on the large peg, then all three blanks in ϵ are filled. Nevertheless, no definite judgment emerges from ϵ in this way. For this purpose it is necessary that the two blanks x and y in ϵ or the objects which fill them be related in a definite way to the two blanks XY of the relation R $(x\,y)$ which fills the blank Z in ϵ or, as I prefer to say, the latter blanks must be "connected" with the former. Accordingly, to the scheme of the relation ϵ there belong two "connection wires" originating from the base of the pegs xy with whose help the connection of the "secondary" blanks to the "primary" ones xy, in the manner indicated by the figure, has to occur in the process of filling the blanks. Clearly, in the case of the given relation R, this connection can occur in two different ways. The fact that the wires are fed into the pegs XY from above is meant to indicate further that all blanks are "saturated" in the filling process. The existence of such "connection wires" in the scheme of a relation carries over naturally from ϵ to relations which are produced from ϵ and the primitive relations by means of the construction principles; the "secondary" blanks of the relations employed in the filling process must be connected by the wires in part to the primary blanks and in part to certain relational points. Some effort is required to understand how the scheme of a relation appears in general and what the process of filling the blanks, which produces a definite judgement scheme, amounts to. The model we have just employed will stand us in good stead here. But even if it is of great value to us in acquiring a comprehensive view of the syntax of relations, this rather involved schematism can be avoided by means of a simple and purely formal device, namely by introducing *subject-ordered* relations in place of relations. Altering somewhat the explanation given in my treatise,[3] I here understand a subject-ordered relation to be one in whose scheme a definite order of succession is established within each group of blanks affiliated with one and the same category of object. If the blanks x and y in the above-mentioned relation ϵ are numbered in this way and if, further, a *subject-ordered binary* relation is always used to fill the blank Z, i.e., if the secondary blanks X and Y are likewise numbered, then the "connection" is superfluous, since it is self-evident that the secondary blank 1 is to be connected to the primary blank 1 and the secondary blank 2 to the primary blank 2. Because of this, ϵ and the relations to be derived from it are of the same sort, as I assumed from the very beginning of my treatise: in order to obtain a meaningful assertion from them, it is sufficient to fill each blank with an object of the relevant category (an object which may itself be a subject-ordered relation). I am even less hesitant to introduce subject-ordered relations, in spite of the formal and artifical character of this device, in view of the fact that, later on, if the transition from relations to sets is to be carried out, one must necessarily assume that relations are subject-ordered.

A quinary relation $R\,(uv\,|\,xyz)$ (to pick a definite example) can also be regarded as a *binary* relation holding between u and v and depending on the three "*arguments*" xyz. It can then be used in a higher level relation to fill a blank Z affiliated with binary relations—in which case only the first two of the secondary blanks *uv* | *xyz* are saturated through connection, while *xyz* remain free blanks which wait to be filled by objects. This process is employed in the principle of substitution 7 (p. 35). If we are just interested in relations between objects of the basic categories, then clearly the introduction of ∈ and the principle of substitution by itself leads to no relations other than those which can be constructed with the help of the first six principles alone, without this expansion. ∈ and the principle of substitution become fruitful only when supplemented by the principle of iteration (8) of which they are the indispensable precursors. Iteration, however, is of the greatest significance for all mathematical concept-formation.

I have called those relations delimited which can be constructed, by the indicated means, from the intuitively exhibited, primitive relations of the underlying sphere of operation. In this way, we obtain the desired, extensionally determinate restriction of the concept "relation" which is required for founding a non-circular version of analysis. Only now, at the end, should the *concept of set and function* be introduced. And, of course, the sets and functions correspond to the (subject-ordered, delimited) relations in such a way that *extension* rather than *sense* determines their equality or distinctness; they form the mathematical superstructure over the intuitively grounded substructure of the basic categories.

As far as I can see, analysis provides no occasion for iterating this expanded mathematical process which includes the principles of construction 7 and 8.[4] (Cf. p. 22 concerning the expression "mathematical process.") And, furthermore, the schema we have secured proves to be comprehensive enough with regard to applications that on this basis a rational *theory of the continuum* can be constructed (as in Chapter 2).

Notes to Chapter 1

1. [I gather that the "basic components" Weyl has in mind are the informal analogues of relation symbols (including, of course, unary ones) and individual constants. His point is that a sentence-form such as 'Fc' or 'c is F' can accomodate both meaningful and meaningless expressions: for example, "This leaf is green" and "Honesty is green."–Trans.]

2. [The paradox which follows was, in fact, devised by Kurt Grelling. But Weyl is right to suggest that it is closely related to Russell's paradox.–Trans.]

3. Formal logicians would be happy to forget that most concepts are inexact (and, indeed, are so by nature–which need not be considered a defect) and that their scope is fluid. Cf. Husserl (1913b: 136ff.). In mathematics, we deal only with entities whose characteristics can be precisely specified.

4. Were one to employ things other than *words* as symbols, such an ordering could perhaps be avoided. For example, let the judgment scheme of a relation be represented by a wooden plate supplied with one peg for each blank of the scheme. And let the relata be represented by little balls, each equipped with a hole so that they can be stuck onto the pegs (to represent the "filling" of the blanks). These are symbols which "in themselves" are just as useful as the words. We can regard properties as relations of a sort (namely, as those whose judgment schemes possess only one blank) in the same way that (contrary to the practice of the Greeks) we consider 1 to be a number.

5. Naturally, whether we are actually able to answer this question is beside the point.

6. We shall eventually refer to these principles simply by citing their numbers.

7. [Evidently, Weyl here assumes that the variables x,y are to range only over *males*. Otherwise, "y is a nephew of x" would not be equivalent to "x is an *uncle* of y."–Trans.]

8. [That is, first, coincidence of blanks is to be marked by filling coinciding (and *only* coinciding) blanks with similar variables; and, second, all the combinatorial possiblities for coincidence are allowed.–Trans.]

9. [Apparently, Weyl uses "$U(xy*) = V(xy)$" to indicate that we are introducing a new relation symbol V into our language and that $V(xy)$ is to be equivalent to $U(xy*)$.–Trans.]

10. [Note that each occurrence of the asterisk signifies a distinct existential quantifier. As the Introduction points out, this will create problems for Weyl.–Trans.]

11. [$W(x*)$ means that $\exists y(U(xy) \cdot V(xy))$; and, as Weyl notes, this is certainly not equivalent to $U(x*) \cdot V(x*)$, which means that $\exists y U(xy) \cdot \exists y V(xy)$. However, $U(x*) \cdot V(x*)$ is equivalent to $T(x**)$, if $T(xyz)$ is equivalent to $U(xy) \cdot V(xz)$.–Trans.]

12. These closed judgment schemes are, intrinsically, just *propositions*. That they all have one meaning, i.e, express a judgment, is a precise formulation of the hypothesis mentioned at the end of §1 regarding the "complete system of self-existent objects."

13. $\overline{[B}(*)$ means that $\exists x–B(x)$. $\overline{B(*)}$ means that $–\exists x\, B(x)$. —Trans.]

14. From now on we shall use the term "judgment" exclusively in its proper sense and no longer for judgment schemes which contain blanks.

15. Alternatively, we could consider particular judgments to be just those whose construction involves no application at all of Pr. 6; but then, in addition to the "particular" and "general" judgments, there would be mixed "general-particular" ones.

16. At least it seems so to me; although some might be of a different opinion even about this.

17. The relation of this *conceptual* homogeneity to the *intuitive* homogeneity of space will not be discussed here.

18. [Note that while $U \cdot V(a)$ evidently means "$U(a) \cdot V(a)$," $U \cdot V(*)$ does not mean "$U(*) \cdot V(*)$" (i.e., "$\exists x U(x) \cdot \exists x V(x)$") but rather "$\exists x(U(x) \cdot V(x))$." So $\overline{U \cdot \overline{V}}(*)$ means "$–\exists x(U(x) \cdot –V(x))$." —Trans.]

19. In the Preface to Dedekind (1888) we read that, "In science, whatever is provable must not be believed without proof." This remark is certainly characteristic of the way most mathematicians think. Nevertheless, it is a preposterous principle. As if such an indirect concatenation of grounds, call it a proof though we may, can awaken any "belief" apart from our assuring ourselves, through immediate insight, that each individual step is correct! In all cases, this process of confirmation (and not the proof) remains the ultimate source from which knowledge derives its authority; it is the "experience of truth." Whoever approaches other disciplines, such as philosophy, in the manner of a mathematician, demanding mathematical definitions and deductions, proceeds no more sagaciously than a zoologist who rejects numbers on the ground that they are not living beings.

20. [For a more detailed discussion of axiomatic systems and their completeness or incompleteness, the reader should consult Weyl (1949a: 18-29). Naturally, Weyl admits that some interesting axiomatic systems are *formally* complete. But he denies that such systems supply a complete account of their intended subject matter.–Trans.]

21. I readily acknowledge that Dedekind's treatise has played an extremely important role in the development of mathematical thought.

22. [Actually, Richard uses a diagonalization procedure to show that "the totality of all numbers defined by finitely many words" apparently fails to contain some finitely defined numbers. See Richard (1905).–Trans.]

23. [Weyl's point is that the infinitude of a set M does not guarantee that there is a *constructive* pairing function whose domain is the set of natural numbers and whose range is a subset of M.–Trans.]

24. The primary significance of the narrower procedure is most clearly conveyed by the following observation: The objects of the basic categories remain uninterruptedly the genuine objects of our investigation only when we comply with the narrower procedure; otherwise, the profusion of derived properties and relations becomes just as much an object of our thought as the realm of those primitive objects. In order to reach a decision about "delimited" judgments, i.e., those which are formed under the restrictions of the narrower procedure, we need only survey these basic objects; "non-delimited" judgments require that one also survey all derived properties and relations.

25. [It is important to understand that Weyl's "narrower procedure" does *not* involve a general restriction on the formation rules of his language. The restriction on the domain of quantification (and on the sorts of individual constants used to fill blanks) applies only to judgment schemes which are being employed to *define sets and functional connections*. In the previous note, Weyl is *not* suggesting that we should, in general, restrict ourselves to "delimited judgments." He is just using the notion of a "delimited judgment" to explain why we should restrict ourselves to "delimited sets." (On our use of the words "delimited judgments" to translate "*finite Urteile*," see note 37.)–Trans.]

26. The construction of the concept of number will be carried out more thoroughly and systematically in Chapter 2.

27. Here I regard the natural numbers as *sui generis*, i.e., as not included in the category containing the rational numbers.

28. [Cf. note 25.–Trans.]

29. In science there are no "commandments", just "laws". So here too, using the term "there is" in connection with objects which do not belong to the basic categories should certainly not be "forbidden". It is, of course, entirely possible (and permissible) to adopt the broader procedure; but if this is done, let it be done in a non-circular way. [There is, in fact, no shortage of constructivists who have successfully adopted the "broader procedure". Cf. Wang (1970: 639-46).–Trans.]

30. One should take care not to introduce the relation between x, y, z and U expressed by $U=F(xyz)$. Otherwise, one would trap oneself in the circle from which we have just now escaped. [Weyl's point is that the principle of substitution is to be used to *eliminate* "higher level" variables such as U.

So one should not then turn around and *reintroduce* them.–Trans.]

31. [Evidently, "$\in(xzX)\cdot_{\mathscr{D}}(xzy)|_{z=_*}$" means that $\exists\, z(\in(xzX)\cdot_{\mathscr{A}}(xzy))$. For an explanation of why Weyl needed to introduce this new notation, see the Introduction.–Trans.]

32. Clearly, Dedekind's definition of infinitude (i.e., that a set is infinite if and only if there is a pairing between it and a proper subset of itself) is not to be entertained here.

33. In any case, it is possible to develop this position in a *logically* complete way. I choose not to discus here whether the concept of cardinal number might not be epistemologically primary and independent of the concept of ordinal number.

34. [For, as mentioned in note 30, the principle of substitution allows us to eliminate occurrences of "higher level" variables. –Trans.]

35. At this point in the production process, the introduction of sets is just a way of allowing for the fact that relations between objects are themselves objects–between which new relations can obtain.

36. [For a discussion of this restriction, see the Introduction.–Trans.]

37. [The German is, ". . . *wo ein kurzes Wort erwünscht ist, nenne ich sie 'finite' Relationen.*" Clearly, we must not translate the German "finite" by the English "finite." (Finite sets are those with only finitely many members–and that is *not* what Weyl is talking about.) The words "finitary" and "finitistic" are not appropriate either, for they already have a technical meaning within Hilbert's formalist program which is very different from what Weyl intends here. "Definite" is better, but it will be useful for other purposes. So we have chosen the word "delimited" which is unusual enough to be recognizable as a technical term and which suggests the restrictions involved in the "narrower procedure."-Trans.]

38. Unless the given set consists of elements of a basic category.

39. As far as I know, this principle of definition was first set forth in Frege (1884: §§63–68)–with clarity greater than that achieved by any later author and with complete awareness of the great significance of this mode of definition for the whole of mathematics. [For a more satisfactory discussion of definition by abstraction, see Weyl (1949a: 8-13).–Trans.]

40. See Russell (1908) or Whitehead and Russell (1910).

41. Poincaré (1905), (1906) and (1909).

42. Zermelo (1908).

43. Ibid., p. 263.

44. Cf. Weyl (1910) (my inaugural lecture).

45. Likewise, of course, the theory of finite sets proposed in Zermelo (1909).

Notes to Chapter 2

1. Cf. Dedekind (1888; §§7, 11, 12).

2. [In the original text, "x" and "x_0" are reversed by mistake. −Trans.]

3. Here and throughout what follows, the denotation of the word "set" is restricted to *delimited* ones. Cf. Chapter 1, §8.

4. I do not consider it a mere historical accident that the fractions appear before the negative numbers. The development of our system accords with this historical course of events by passing from the *fractions* (not from the *whole* numbers) to the rationals.

5. Only in a thoroughly figurative sense can one speak, as is done nowadays, of a function of infinitely many variables $f(1)$, $f(2)$, $f(3)$. . . .

6. We have already given this definition (in Chapter 1, §2, where it served as a way of illustrating our symbolism for relations). We repeat it here because it is important for us to require, from the very start, that the right-hand sides a, b of the characteristic inequalities be fractions rather than positive real numbers.

7. Of course, in the case of every function one encounters in analysis, this question does *not* remain open, since the negative judgment which asserts their continuity is a logical consequence of the "axioms" into which the principles of definition change when formulated as positive existential judgments concerning sets. But this is just a special characteristic of these "absolutely" continuous functions.

8. For example, see the opening pages of Bergson (1907).

9. Do not forget that in the "continuum" of the real numbers, the individual elements are in fact *precisely* as isolated from one another as, say, the whole numbers.

10. Concerning the problem of time, cf. Husserl (1913b, §§81, 82) and Linke (1912: Ch. 6).

11. For example, our inability to connect up the continuous with the schema of the whole numbers is not just a matter of personal preference. But who knows what the future and quantum mechanics have in store in the physical domain?

12. Of course, this applies only to the relations of our sphere of operation, among which, e.g., the non-intersection of two spans is not included.

13. In particular, see Chapter 1, §4, "Begriff der Fläche." Also cf. Hausdorff (1914), Chapters 7 and 8, above all p. 213.

14. [The pre-critical notions have, in fact, been revived in modern non-standard analysis−whose consistency relative to standard analysis is

guaranteed by Gödel's compactness theorem. Cf. Robinson (1966).–Trans.]

15. What "continuous" means here will be immediately understood.

Notes to Appendix

1. [This is a translation of Weyl (1919).–Trans.]

2. Concerning the idea of a "pure grammar" cf. Husserl (1913a, 2:328 ff.)

3. [On p. 22, Weyl required that all the blanks of a scheme be arranged in a single order of succession.–Trans.]

4. [The "mathematical process" transforms intensional entities into extensional ones–e.g. relations into sets. Weyl's point here is, first, that the delimited sets can be introduced "all at once" through a single application of the "mathematical process" and, second, that there is then no need to construct new relations which involve these sets and which are themselves transformed into sets by a further application of the "mathematical process." Such further applications are what Weyl means by the "iteration of the mathematical process," which is not to be confused with the iteration involved in applications of principle 8.–Trans.]

Bibliography

In addition to those cited above, this bibliography lists other works which are likely to be of interest to students of Weyl's philosophy of mathematics.

Beisswanger, P.
    1965   *Die Anfechtbarkeit der klassischen Mathematik: Studien über Hermann Weyl.* Stuttgart dissertation.
    1966   "Hermann Weyl and Mathematical Texts," *Ratio* 8: 25-45.

Bergson, H.
    1907   *L'Evolution Créatrice.* Paris: Alcan. English translation = Bergson (1911).
    1911   *Creative Evolution.* Translated by A. Mitchell. London: Macmillan.

Dedekind, R.
    1888   *Was sind und was sollen die Zahlen?* Brunswick: Vieweg. English translation in Dedekind (1963).
    1963   *Essays on the Theory of Numbers.* Translated by W. W. Beman. New York: Dover.

Fichte, J. G.
    1912   *Werke.* Edited by F. Medicus. Leipzig: Meiner. Vol. 6.

Frege, G.
    1884   *Die Grundlagen der Arithmetik.* Breslau: Koebner. English translation = Frege (1980).
    1893   *Grundgesetze der Arithmetik.* Jena: Pohle. Partial English translation in Frege (1964).
    1964   *Basic Laws of Arithmetic: Exposition of the System.* Translated and edited, with an introduction, by M. Furth. Berkeley: University of California Press.
    1980   *The Foundations of Arithmetic.* Translated by J. L. Austin. Evanston, Il.: Northwestern University Press.

Grebe, W.
    1920   *Beiträge zum Aufbau der Mengenlehre auf Grund vom Weyls Theorie des Kontinuums.* Frankfurt dissertation.

Hausdorff, F.
    1914   *Grundzöge der Mengenlehre.* Leipzig: Veit; reprinted New

York: Chelsea, 1949. English translation = Hausdorff (1957).

1957   *Set Theory*. Translated by J. R. Aumann. New York: Chelsea.

Hölder, O.

1892   "Review: R. Grassmann's *Zahlenlehre*," *Göttinger gelehrten Anzeigen*.

1926   "Der angebliche circulus vitiosus und die sogenannte Grundlagenkrise in der Analysis," *Berichte über die Verhandlungen der Sächsischen Akademie der Wissenschaften zu Leipzig. Mathematische-physische Klasse*, 78: 243-50.

Husserl, E.

1913a   *Logische Untersuchungen*, 2d ed. Halle: Niemeyer; reprinted in Husserl (1984). English translation = Husserl (1970).

1913b   "Ideen zu einer reinen Phänomenologie und phänomenologischen Philosophie," *Jahrbuch für Philosophie und Phänomenologische Forschung* 1; reprinted in Husserl (1950). English translation = Husserl (1962).

1950   *Husserliana*. Edited by W. Biemal. The Hague: Nijhoff. Vol. 3.

1962   *Ideas*. Translated by W. R. B. Gibson. New York: Collier Books.

1970   *Logical Investigations*. Translated by J. N. Findlay. London: Routledge & Kegan Paul.

1984   *Husserliana*. The Hague: Nijhoff. Vol. 19.

Linke, P. F.

1912   *Die phänomenale Sphäre und das reale Bewusstsein*. Halle: Niemeyer.

Lorenzen, P.

1950   "Konstruktive Begründung der Mathematik," *Mathematische Zeitschrift* 53: 162-201.

1951   "Die Widerspruchfreiheit der klassischen Analysis," *Mathematische Zeitschrift* 54: 1-24.

1955   *Einführung in die operative Logik und Mathematik*. Berlin: Springer.

1958   "Logical Reflection and Formalism," *Journal of Symbolic Logic* 23: 241-49.

1962a   *Metamathematik*. Mannheim: Bibliographisches Institut.

1962b   "Equality and Abstraction," *Ratio* 4, 85-90.

1965   *Formal Logic*. Translated by F. J. Crosson. Dordrecht-Holland: D. Reidel.

1971   *Differential and Integral*. Translated by J. Bacon. Austin: University of Texas Press.

Poincaré, H.

1905   "Les mathématiques et la logique," *Revue de Métaphysique et de Morale* 13: 815-35.

1906   "Les mathématiques et la logique," *Revue de Métaphysique et de Morale* 14: 17-34, 294-317.

1909   "Réflexions sur les Deux Notes Précédentes," *Acta Mathematica* 32: 198-200.

1913   *The Foundations of Science.* Translated by G. B. Halstead New York: The Science Press.

Richard, J.

1905   "Les Principes des Mathématiques et le Problème des Ensembles," *Revue Générale des Sciences Pures et Appliquées* 16: 541. English translation in van Heijenoort (1967).

Robinson, A.

1966   *Non-Standard Analysis.* Amsterdam: North Holland.

Russell, B.

1908   "Mathematical Logic as Based on the Theory of Types," *American Journal of Mathematics* 30: 222-62. Reprinted in van Heijenoort (1967).

van Dalen, D.

1984   "Four Letters from Edmund Husserl to Hermann Weyl," *Husserl Studies* 1 (no.1).

van Heijenoort, J.

1967   *From Frege to Gödel: A Source Book in Mathematical Logic, 1879-1931.* Cambridge: Harvard University Press.

Wang, H.

1970   *Logic, Computers, and Sets.* New York: Chelsea.

Weyl, H.

1910   "Über die Definitionen der mathematischen Grundbegriffe," *Mathematisch-naturwissenschaftliche Blätter* 7, 93-95, 109-13.

1913   *Die Idee der riemannschen Fläche* Leipzig: Teubner; English translation = Weyl (1955).

1919   "Der circulus vitiosus in der heutigen Begründung der Analysis," *Jahresbericht der Deutschen Mathematikervereinigung* 28, 85-92.

1921   "Über die neue Grundlagenkrise der Mathematik," *Mathematische Zeitschrift* 10, 39-79.

1924   "Randbemerkungen zu Hauptproblemen der Mathematik," *Mathematische Zeitschrift* 20, 131-50.

1925   "Die heutige Erkenntnislage in der Mathematik," *Symposion* 1: 1-32.

1928   "Diskussionbemerkungen zu dem zweiten Hilbertschen Vortrag über die Grundlagen der Mathematik," *Abhandlungen aus dem mathematischen Seminar der Hamburgischen Universität* 6: 86-88; English translation in van Heijenoort (1967).

1929   "Consistency in Mathematics," *The Rice Institute Pamphlet* 16: 245-65.

1931   *Die Stufen des Unendlichen.* Jena: Fischer.

1940a   "The Ghost of Modality," *Philosophical Essays in Memory of Edmund Husserl.* Cambridge: Harvard University Press, 278-303.

1940b   "The Mathematical Way of Thinking," *Science* 92: 437-46.

1944a   "Obituary: David Hilbert 1862-1943," *Obituary Notices of Fellows of the Royal Society* 4: 547-53.

1944b   "David Hilbert and his Mathematical Work," *Bulletin of the American Mathematical Society* 50, 612-54.

1946a   "Mathematics and Logic. A brief survey serving as a preface to a review of 'The Philosophy of Bertrand Russell,'" *The American Mathematical Monthly* 53: 2-13.

1946b   "Review: The Philosophy of Bertrand Russell," *The American Mathematical Monthly* 53: 208-14.

1949a   *Philosophy of Mathematics and Natural Science.* Princeton: Princeton University Press.

1949b   "Wissenschaft als symbolische Konstruktion des Menschen," *Eranos-Jahrbuch* 16, 375-431.

1953   "Über den Symbolismus der Mathematik und mathematischen Physik," *Studium generale* 6: 219-28.

1954a   "Address on the unity of knowledge delivered at the bicentennial conference of Columbia University," Columbia University in the City of New York Bicentennial Celebration.

1954b   "Erkenntnis und Besinnung," *Studia Philosophica, Jahrbuch der schweizerischen philosophischen Gesellschaft; Annuaire de la Société Suisse de Philosophie.*

1955   *The Concept of a Riemann Surface.* Translated by G. R. Maclane. Reading, Mass.: Addison-Wesley.

1968   *Gesammelte Abhandlungen.* Edited by K. Chandrasekharan. Berlin: Springer.

Whitehead, A. N. and B. Russell

1910   *Principia Mathematica.* Cambridge: Cambridge University Press. Vol. 1.

Zermelo, E.

1908   "Untersuchungen über die Grundlagen der Mengenlehre I," *Mathematische Annalen* 65: 261-81. English translation in van Heijenoort (1967).

1909   "Sur les Ensembles Finis et le Principe de l'Induction Complète," *Acta Mathematica* 32: 185-93.

Index

absolute sphere of operation, 28
absurdity, 6,16
addition, 51-53, 59, 68, 99
algebra: algebraic number, 72-74; algebraic principles of construction, 46, 70; fundamental theorem of algebra, 85
applications, 32
Archimedean Axiom, 100
arithmetic, 15-16, 51-54
associative law, 52, 54
attributable properties, 20
autological words, 6
axioms, 17-19, 44, 48, 122; Hibert's axiom of completeness, 95; "Axiom of Reducibility," 47

Bergson, 90, 122

Cantor, 23, 24, 27, 47, 48, 86
cardinal number, 24, 39, 55, 121; cardinality, 38-39, 55-57
category, 5-9, 34; basic categories, 28, 30, 41, 48, 68
causal connections, 46
circulus vitiosus of analysis, 32, 48, 109
coinciding blanks, 10
commutative law, 52, 54
completeness of theories, 14, 19, 119
complete induction, 25, 39
complex numbers, 74
consistency, 19, 44
continuity, 14, 24, 80-86, 92-96
convergence, 75; Cauchy's convergence principle, 34, 75, 90, 95
counting, process of, 57
cut, 67; of the number sequence, 54; open cut, 67

Dedekind, 23, 24, 31, 45, 47, 48, 111, 119, 121, 122; Dedekind's cut principle, 77, 90
deductive method, 17
definition by abstraction, 45, 121
delimited properties and relations, 43, 81, 87, 120, 121
denumerability, 26-28
dependent blanks, 34
differentiation, 86
Dirichlet's Principle, 79
distributive law, 53

domain of fractions, 62

equivalence, 16
existence, 8, 41, 49, 109-10
extension, 8, 110, 117; extensionally determinate, 109-11; extensionally equivalent, 110

"Fermat's last theorem", 18, 49
Fichte, 7, 111
"filling in", 7, 10
"finite definition", 27
formalism, 1, 2, 6, 18
fractions, 58-63, 122
Frege, 2, 47, 121
function, 23, 33-36, 43, 45-46, 58, 114, 117; continuous functions, 81-86; elementary functions, 80; function of, 33, 34, 50; function sequences, 76; "number-theoretic functions," 58; real-valued function, 67
functional connection, 21-22

Galileo, 46, 113

Heine-Borel theorem, 77-78
heterological words, 6
homogeneity, 16, 98
Husserl, 2, 111, 118, 122, 123
hyperanalysis, 95

ideal(s), 45; ideal elements, introduction of, 45
identity, 9, 42
independent blanks, 34
individuals, 15
inference, forms of, 39
infinity, 24, 49, 121; infinite series, 79
insight, 18, 19, 111
integration, 86
iteration, principle of, 26, 35-39, 42; concept of iteration, 19, 48; narrower iteration procedure, 30-32, 120

judgment, 5; closed judgment, 10; existential judgments, 8; general judgment, 14, 17; judgment schemes, 7; complex judgment schemes, 9; primitive judgment schemes, 9; simple judgment schemes, 9; particular judgments, 14; pertinent judgments, 11, 17, 43; simple judgments, 9; true judgments, 17

129

law of nature, 46
least upper bound, 31, 83, 112
levels, 40; first level, 28-30, 40; second level, 29-30, 40
limit inferior, 75
logical consequence, 16, 18
logical principles of construction, 46, 70, 113
logical structure, 6, 12, 16

mathematical: mathematical principle (iteration), 38; mathematical process, 22, 28, 40
mathesis pura, 19
meaning, 5, 111
measure, 26, 99
multi-dimensional set, 22
multiplication, 53, 59, 65, 99; multipliers, 58

natural numbers, 15-16, 24-28, 37, 49, 51-56, 110
nullification, 63

objective correspondence, 23 objective fact, 20
open remainder, 69
order of succession, 8
ordinal numbers, 24, 121

Poincaré, 47, 48, 121
power series, 80
projection, 22, 118
proof, 17, 119
properties, 5, 20, 23, 33, 110, 118; derived properties, 11, 20, 120 primitive properties, 206
propositions, 5
pure number theory, 26
Pythagoras 1, 24

rational numbers, 30, 66-72, 111; domain of, 66; negative rational numbers, 64; positive rational numbers, 64
real number, 31, 66-72, 111; negative real numbers, 69; positive real numbers, 69; sequence of, 67
relations, 7, 33; derived relations, 11, 22, 120; first level relations, 29; primitive relations, 22, 41
replication, 58, 61, 65, 89
reversal, 63, 65, 69
Richard's antimony, 26-27, 120

Russell, 6, 29, 47, 121

self-evident, 16
sets, 20-24, 25, 28-30, 40-45, 49, 114, 117; two-, three-, four-dimensional sets, 21; sphere of existence, 116
sphere of operation, 24, 25, 28, 44, 122
state of affairs, 5
stipulations, 18
subdivision, 58, 61, 65, 69
subject ordered: subject-ordered judgment schemes, 22; subject-ordered relations, 22, 116
substitution, principle of, 35, 42, 117; substitution of elements, 56,

theory of the continuum, 93, 117
time, 46, 88-94
truth, 5, 6

ultimate principles, 23
uniform continuity, 13, 81-82
universal: universal scale of infinite cardinal and ordinal nu;mbers, 24; universal set theory, 24

value range, 47, 81
variables, 7, 33, 34
"vicious-circle" principle, 47

Weierstrass, 23, 79

Zermelo, 48, 121; "Subset"-Axiom III, 48

A CATALOG OF SELECTED
DOVER BOOKS
IN SCIENCE AND MATHEMATICS

A CATALOG OF SELECTED
DOVER BOOKS
IN SCIENCE AND MATHEMATICS

QUALITATIVE THEORY OF DIFFERENTIAL EQUATIONS, V.V. Nemytskii and V.V. Stepanov. Classic graduate-level text by two prominent Soviet mathematicians covers classical differential equations as well as topological dynamics and ergodic theory. Bibliographies. 523pp. 5⅜ × 8½. 65954-2 Pa. $10.95

MATRICES AND LINEAR ALGEBRA, Hans Schneider and George Phillip Barker. Basic textbook covers theory of matrices and its applications to systems of linear equations and related topics such as determinants, eigenvalues and differential equations. Numerous exercises. 432pp. 5⅜ × 8½. 66014-1 Pa. $9.95

QUANTUM THEORY, David Bohm. This advanced undergraduate-level text presents the quantum theory in terms of qualitative and imaginative concepts, followed by specific applications worked out in mathematical detail. Preface. Index. 655pp. 5⅜ × 8½. 65969-0 Pa. $13.95

ATOMIC PHYSICS (8th edition), Max Born. Nobel laureate's lucid treatment of kinetic theory of gases, elementary particles, nuclear atom, wave-corpuscles, atomic structure and spectral lines, much more. Over 40 appendices, bibliography. 495pp. 5⅜ × 8½. 65984-4 Pa. $12.95

ELECTRONIC STRUCTURE AND THE PROPERTIES OF SOLIDS: The Physics of the Chemical Bond, Walter A. Harrison. Innovative text offers basic understanding of the electronic structure of covalent and ionic solids, simple metals, transition metals and their compounds. Problems. 1980 edition. 582pp. 6⅛ × 9¼. 66021-4 Pa. $15.95

BOUNDARY VALUE PROBLEMS OF HEAT CONDUCTION, M. Necati Özisik. Systematic, comprehensive treatment of modern mathematical methods of solving problems in heat conduction and diffusion. Numerous examples and problems. Selected references. Appendices. 505pp. 5⅜ × 8½. 65990-9 Pa. $11.95

A SHORT HISTORY OF CHEMISTRY (3rd edition), J.R. Partington. Classic exposition explores origins of chemistry, alchemy, early medical chemistry, nature of atmosphere, theory of valency, laws and structure of atomic theory, much more. 428pp. 5⅜ × 8½. (Available in U.S. only) 65977-1 Pa. $10.95

A HISTORY OF ASTRONOMY, A. Pannekoek. Well-balanced, carefully reasoned study covers such topics as Ptolemaic theory, work of Copernicus, Kepler, Newton, Eddington's work on stars, much more. Illustrated. References. 521pp. 5⅜ × 8½. 65994-1 Pa. $12.95

PRINCIPLES OF METEOROLOGICAL ANALYSIS, Walter J. Saucier. Highly respected, abundantly illustrated classic reviews atmospheric variables, hydrostatics, static stability, various analyses (scalar, cross-section, isobaric, isentropic, more). For intermediate meteorology students. 454pp. 6⅛ × 9¼. 65979-8 Pa. $14.95

RELATIVITY, THERMODYNAMICS AND COSMOLOGY, Richard C. Tolman. Landmark study extends thermodynamics to special, general relativity; also applications of relativistic mechanics, thermodynamics to cosmological models. 501pp. 5⅜ × 8½. 65383-8 Pa. $12.95

APPLIED ANALYSIS, Cornelius Lanczos. Classic work on analysis and design of finite processes for approximating solution of analytical problems. Algebraic equations, matrices, harmonic analysis, quadrature methods, much more. 559pp. 5⅜ × 8½. 65656-X Pa. $12.95

SPECIAL RELATIVITY FOR PHYSICISTS, G. Stephenson and C.W. Kilmister. Concise elegant account for nonspecialists. Lorentz transformation, optical and dynamical applications, more. Bibliography. 108pp. 5⅜ × 8½. 65519-9 Pa. $4.95

INTRODUCTION TO ANALYSIS, Maxwell Rosenlicht. Unusually clear, accessible coverage of set theory, real number system, metric spaces, continuous functions, Riemann integration, multiple integrals, more. Wide range of problems. Undergraduate level. Bibliography. 254pp. 5⅜ × 8½. 65038-3 Pa. $7.95

INTRODUCTION TO QUANTUM MECHANICS With Applications to Chemistry, Linus Pauling & E. Bright Wilson, Jr. Classic undergraduate text by Nobel Prize winner applies quantum mechanics to chemical and physical problems. Numerous tables and figures enhance the text. Chapter bibliographies. Appendices. Index. 468pp. 5⅜ × 8½. 64871-0 Pa. $11.95

ASYMPTOTIC EXPANSIONS OF INTEGRALS, Norman Bleistein & Richard A. Handelsman. Best introduction to important field with applications in a variety of scientific disciplines. New preface. Problems. Diagrams. Tables. Bibliography. Index. 448pp. 5⅜ × 8½. 65082-0 Pa. $12.95

MATHEMATICS APPLIED TO CONTINUUM MECHANICS, Lee A. Segel. Analyzes models of fluid flow and solid deformation. For upper-level math, science and engineering students. 608pp. 5⅜ × 8½. 65369-2 Pa. $13.95

ELEMENTS OF REAL ANALYSIS, David A. Sprecher. Classic text covers fundamental concepts, real number system, point sets, functions of a real variable, Fourier series, much more. Over 500 exercises. 352pp. 5⅜ × 8½. 65385-4 Pa. $10.95

PHYSICAL PRINCIPLES OF THE QUANTUM THEORY, Werner Heisenberg. Nobel Laureate discusses quantum theory, uncertainty, wave mechanics, work of Dirac, Schroedinger, Compton, Wilson, Einstein, etc. 184pp. 5⅜ × 8½.
60113-7 Pa. $5.95

INTRODUCTORY REAL ANALYSIS, A.N. Kolmogorov, S.V. Fomin. Translated by Richard A. Silverman. Self-contained, evenly paced introduction to real and functional analysis. Some 350 problems. 403pp. 5⅜ × 8½. 61226-0 Pa. $9.95

PROBLEMS AND SOLUTIONS IN QUANTUM CHEMISTRY AND PHYSICS, Charles S. Johnson, Jr. and Lee G. Pedersen. Unusually varied problems, detailed solutions in coverage of quantum mechanics, wave mechanics, angular momentum, molecular spectroscopy, scattering theory, more. 280 problems plus 139 supplementary exercises. 430pp. 6½ × 9¼. 65236-X Pa. $12.95

ASYMPTOTIC METHODS IN ANALYSIS, N.G. de Bruijn. An inexpensive, comprehensive guide to asymptotic methods—the pioneering work that teaches by explaining worked examples in detail. Index. 224pp. 5⅜ × 8½.   64221-6 Pa. $6.95

OPTICAL RESONANCE AND TWO-LEVEL ATOMS, L. Allen and J.H. Eberly. Clear, comprehensive introduction to basic principles behind all quantum optical resonance phenomena. 53 illustrations. Preface. Index. 256pp. 5⅜ × 8½.
65533-4 Pa. $7.95

COMPLEX VARIABLES, Francis J. Flanigan. Unusual approach, delaying complex algebra till harmonic functions have been analyzed from real variable viewpoint. Includes problems with answers. 364pp. 5⅜ × 8½.   61388-7 Pa. $8.95

ATOMIC SPECTRA AND ATOMIC STRUCTURE, Gerhard Herzberg. One of best introductions; especially for specialist in other fields. Treatment is physical rather than mathematical. 80 illustrations. 257pp. 5⅜ × 8½.   60115-3 Pa. $5.95

APPLIED COMPLEX VARIABLES, John W. Dettman. Step-by-step coverage of fundamentals of analytic function theory—plus lucid exposition of five important applications: Potential Theory; Ordinary Differential Equations; Fourier Transforms; Laplace Transforms; Asymptotic Expansions. 66 figures. Exercises at chapter ends. 512pp. 5⅜ × 8½.   64670-X Pa. $11.95

ULTRASONIC ABSORPTION: An Introduction to the Theory of Sound Absorption and Dispersion in Gases, Liquids and Solids, A.B. Bhatia. Standard reference in the field provides a clear, systematically organized introductory review of fundamental concepts for advanced graduate students, research workers. Numerous diagrams. Bibliography. 440pp. 5⅜ × 8½.   64917-2 Pa. $11.95

UNBOUNDED LINEAR OPERATORS: Theory and Applications, Seymour Goldberg. Classic presents systematic treatment of the theory of unbounded linear operators in normed linear spaces with applications to differential equations. Bibliography. 199pp. 5⅜ × 8½.   64830-3 Pa. $7.95

LIGHT SCATTERING BY SMALL PARTICLES, H.C. van de Hulst. Comprehensive treatment including full range of useful approximation methods for researchers in chemistry, meteorology and astronomy. 44 illustrations. 470pp. 5⅜ × 8½.   64228-3 Pa. $10.95

CONFORMAL MAPPING ON RIEMANN SURFACES, Harvey Cohn. Lucid, insightful book presents ideal coverage of subject. 334 exercises make book perfect for self-study. 55 figures. 352pp. 5⅜ × 8¼.   64025-6 Pa. $9.95

OPTICKS, Sir Isaac Newton. Newton's own experiments with spectroscopy, colors, lenses, reflection, refraction, etc., in language the layman can follow. Foreword by Albert Einstein. 532pp. 5⅜ × 8½.   60205-2 Pa. $9.95

GENERALIZED INTEGRAL TRANSFORMATIONS, A.H. Zemanian. Graduate-level study of recent generalizations of the Laplace, Mellin, Hankel, K. Weierstrass, convolution and other simple transformations. Bibliography. 320pp. 5⅜ × 8½.   65375-7 Pa. $8.95

THE ELECTROMAGNETIC FIELD, Albert Shadowitz. Comprehensive under-graduate text covers basics of electric and magnetic fields, builds up to electromagnetic theory. Also related topics, including relativity. Over 900 problems. 768pp. 5⅜ × 8¼. 65660-8 Pa. $18.95

FOURIER SERIES, Georgi P. Tolstov. Translated by Richard A. Silverman. A valuable addition to the literature on the subject, moving clearly from subject to subject and theorem to theorem. 107 problems, answers. 336pp. 5⅜ × 8½.
63317-9 Pa. $8.95

THEORY OF ELECTROMAGNETIC WAVE PROPAGATION, Charles Herach Papas. Graduate-level study discusses the Maxwell field equations, radiation from wire antennas, the Doppler effect and more. xiii + 244pp. 5⅜ × 8½.
65678-0 Pa. $6.95

DISTRIBUTION THEORY AND TRANSFORM ANALYSIS: An Introduction to Generalized Functions, with Applications, A.H. Zemanian. Provides basics of distribution theory, describes generalized Fourier and Laplace transformations. Numerous problems. 384pp. 5⅜ × 8½. 65479-6 Pa. $9.95

THE PHYSICS OF WAVES, William C. Elmore and Mark A. Heald. Unique overview of classical wave theory. Acoustics, optics, electromagnetic radiation, more. Ideal as classroom text or for self-study. Problems. 477pp. 5⅜ × 8½.
64926-1 Pa. $12.95

CALCULUS OF VARIATIONS WITH APPLICATIONS, George M. Ewing. Applications-oriented introduction to variational theory develops insight and promotes understanding of specialized books, research papers. Suitable for advanced undergraduate/graduate students as primary, supplementary text. 352pp. 5⅜ × 8½. 64856-7 Pa. $8.95

A TREATISE ON ELECTRICITY AND MAGNETISM, James Clerk Maxwell. Important foundation work of modern physics. Brings to final form Maxwell's theory of electromagnetism and rigorously derives his general equations of field theory. 1,084pp. 5⅜ × 8½. 60636-8, 60637-6 Pa., Two-vol. set $19.90

AN INTRODUCTION TO THE CALCULUS OF VARIATIONS, Charles Fox. Graduate-level text covers variations of an integral, isoperimetrical problems, least action, special relativity, approximations, more. References. 279pp. 5⅜ × 8½.
65499-0 Pa. $7.95

HYDRODYNAMIC AND HYDROMAGNETIC STABILITY, S. Chandrasekhar. Lucid examination of the Rayleigh-Benard problem; clear coverage of the theory of instabilities causing convection. 704pp. 5⅜ × 8¼. 64071-X Pa. $14.95

CALCULUS OF VARIATIONS, Robert Weinstock. Basic introduction covering isoperimetric problems, theory of elasticity, quantum mechanics, electrostatics, etc. Exercises throughout. 326pp. 5⅜ × 8½. 63069-2 Pa. $7.95

DYNAMICS OF FLUIDS IN POROUS MEDIA, Jacob Bear. For advanced students of ground water hydrology, soil mechanics and physics, drainage and irrigation engineering and more. 335 illustrations. Exercises, with answers. 784pp. 6⅛ × 9¼. 65675-6 Pa. $19.95

NUMERICAL METHODS FOR SCIENTISTS AND ENGINEERS, Richard Hamming. Classic text stresses frequency approach in coverage of algorithms, polynomial approximation, Fourier approximation, exponential approximation, other topics. Revised and enlarged 2nd edition. 721pp. 5⅜ × 8½.
65241-6 Pa. $14.95

THEORETICAL SOLID STATE PHYSICS, Vol. I: Perfect Lattices in Equilibrium; Vol. II: Non-Equilibrium and Disorder, William Jones and Norman H. March. Monumental reference work covers fundamental theory of equilibrium properties of perfect crystalline solids, non-equilibrium properties, defects and disordered systems. Appendices. Problems. Preface. Diagrams. Index. Bibliography. Total of 1,301pp. 5⅜ × 8½. Two volumes. Vol. I 65015-4 Pa. $14.95
Vol. II 65016-2 Pa. $14.95

OPTIMIZATION THEORY WITH APPLICATIONS, Donald A. Pierre. Broad-spectrum approach to important topic. Classical theory of minima and maxima, calculus of variations, simplex technique and linear programming, more. Many problems, examples. 640pp. 5⅜ × 8½. 65205-X Pa. $14.95

THE MODERN THEORY OF SOLIDS, Frederick Seitz. First inexpensive edition of classic work on theory of ionic crystals, free-electron theory of metals and semiconductors, molecular binding, much more. 736pp. 5⅜ × 8½.
65482-6 Pa. $15.95

ESSAYS ON THE THEORY OF NUMBERS, Richard Dedekind. Two classic essays by great German mathematician: on the theory of irrational numbers; and on transfinite numbers and properties of natural numbers. 115pp. 5⅜ × 8½.
21010-3 Pa. $4.95

THE FUNCTIONS OF MATHEMATICAL PHYSICS, Harry Hochstadt. Comprehensive treatment of orthogonal polynomials, hypergeometric functions, Hill's equation, much more. Bibliography. Index. 322pp. 5⅜ × 8½. 65214-9 Pa. $9.95

NUMBER THEORY AND ITS HISTORY, Oystein Ore. Unusually clear, accessible introduction covers counting, properties of numbers, prime numbers, much more. Bibliography. 380pp. 5⅜ × 8½. 65620-9 Pa. $9.95

THE VARIATIONAL PRINCIPLES OF MECHANICS, Cornelius Lanczos. Graduate level coverage of calculus of variations, equations of motion, relativistic mechanics, more. First inexpensive paperbound edition of classic treatise. Index. Bibliography. 418pp. 5⅜ × 8½. 65067-7 Pa. $11.95

MATHEMATICAL TABLES AND FORMULAS, Robert D. Carmichael and Edwin R. Smith. Logarithms, sines, tangents, trig functions, powers, roots, reciprocals, exponential and hyperbolic functions, formulas and theorems. 269pp. 5⅜ × 8½. 60111-0 Pa. $6.95

THEORETICAL PHYSICS, Georg Joos, with Ira M. Freeman. Classic overview covers essential math, mechanics, electromagnetic theory, thermodynamics, quantum mechanics, nuclear physics, other topics. First paperback edition. xxiii + 885pp. 5⅜ × 8½. 65227-0 Pa. $19.95

HANDBOOK OF MATHEMATICAL FUNCTIONS WITH FORMULAS, GRAPHS, AND MATHEMATICAL TABLES, edited by Milton Abramowitz and Irene A. Stegun. Vast compendium: 29 sets of tables, some to as high as 20 places. 1,046pp. 8 × 10½. 61272-4 Pa. $24.95

MATHEMATICAL METHODS IN PHYSICS AND ENGINEERING, John W. Dettman. Algebraically based approach to vectors, mapping, diffraction, other topics in applied math. Also generalized functions, analytic function theory, more. Exercises. 448pp. 5⅜ × 8¼. 65649-7 Pa. $9.95

A SURVEY OF NUMERICAL MATHEMATICS, David M. Young and Robert Todd Gregory. Broad self-contained coverage of computer-oriented numerical algorithms for solving various types of mathematical problems in linear algebra, ordinary and partial, differential equations, much more. Exercises. Total of 1,248pp. 5⅜ × 8½. Two volumes. Vol. I 65691-8 Pa. $14.95
Vol. II 65692-6 Pa. $14.95

TENSOR ANALYSIS FOR PHYSICISTS, J.A. Schouten. Concise exposition of the mathematical basis of tensor analysis, integrated with well-chosen physical examples of the theory. Exercises. Index. Bibliography. 289pp. 5⅜ × 8½.
65582-2 Pa. $8.95

INTRODUCTION TO NUMERICAL ANALYSIS (2nd Edition), F.B. Hildebrand. Classic, fundamental treatment covers computation, approximation, interpolation, numerical differentiation and integration, other topics. 150 new problems. 669pp. 5⅜ × 8½. 65363-3 Pa. $14.95

INVESTIGATIONS ON THE THEORY OF THE BROWNIAN MOVEMENT, Albert Einstein. Five papers (1905–8) investigating dynamics of Brownian motion and evolving elementary theory. Notes by R. Fürth. 122pp. 5⅜ × 8½.
60304-0 Pa. $4.95

CATASTROPHE THEORY FOR SCIENTISTS AND ENGINEERS, Robert Gilmore. Advanced-level treatment describes mathematics of theory grounded in the work of Poincaré, R. Thom, other mathematicians. Also important applications to problems in mathematics, physics, chemistry and engineering. 1981 edition. References. 28 tables. 397 black-and-white illustrations. xvii + 666pp. 6⅛ × 9¼.
67539-4 Pa. $16.95

AN INTRODUCTION TO STATISTICAL THERMODYNAMICS, Terrell L. Hill. Excellent basic text offers wide-ranging coverage of quantum statistical mechanics, systems of interacting molecules, quantum statistics, more. 523pp. 5⅜ × 8½. 65242-4 Pa. $12.95

ELEMENTARY DIFFERENTIAL EQUATIONS, William Ted Martin and Eric Reissner. Exceptionally clear, comprehensive introduction at undergraduate level. Nature and origin of differential equations, differential equations of first, second and higher orders. Picard's Theorem, much more. Problems with solutions. 331pp. 5⅜ × 8½. 65024-3 Pa. $8.95

STATISTICAL PHYSICS, Gregory H. Wannier. Classic text combines thermodynamics, statistical mechanics and kinetic theory in one unified presentation of thermal physics. Problems with solutions. Bibliography. 532pp. 5⅜ × 8½.
65401-X Pa. $11.95

ORDINARY DIFFERENTIAL EQUATIONS, Morris Tenenbaum and Harry Pollard. Exhaustive survey of ordinary differential equations for undergraduates in mathematics, engineering, science. Thorough analysis of theorems. Diagrams. Bibliography. Index. 818pp. 5⅜ × 8½. 64940-7 Pa. $16.95

STATISTICAL MECHANICS: Principles and Applications, Terrell L. Hill. Standard text covers fundamentals of statistical mechanics, applications to fluctuation theory, imperfect gases, distribution functions, more. 448pp. 5⅜ × 8½. 65390-0 Pa. $9.95

ORDINARY DIFFERENTIAL EQUATIONS AND STABILITY THEORY: An Introduction, David A. Sánchez. Brief, modern treatment. Linear equation, stability theory for autonomous and nonautonomous systems, etc. 164pp. 5⅜ × 8¼. 63828-6 Pa. $5.95

THIRTY YEARS THAT SHOOK PHYSICS: The Story of Quantum Theory, George Gamow. Lucid, accessible introduction to influential theory of energy and matter. Careful explanations of Dirac's anti-particles, Bohr's model of the atom, much more. 12 plates. Numerous drawings. 240pp. 5⅜ × 8½. 24895-X Pa. $6.95

THEORY OF MATRICES, Sam Perlis. Outstanding text covering rank, non-singularity and inverses in connection with the development of canonical matrices under the relation of equivalence, and without the intervention of determinants. Includes exercises. 237pp. 5⅜ × 8½. 66810-X Pa. $7.95

GREAT EXPERIMENTS IN PHYSICS: Firsthand Accounts from Galileo to Einstein, edited by Morris H. Shamos. 25 crucial discoveries: Newton's laws of motion, Chadwick's study of the neutron, Hertz on electromagnetic waves, more. Original accounts clearly annotated. 370pp. 5⅜ × 8½. 25346-5 Pa. $10.95

INTRODUCTION TO PARTIAL DIFFERENTIAL EQUATIONS WITH APPLICATIONS, E.C. Zachmanoglou and Dale W. Thoe. Essentials of partial differential equations applied to common problems in engineering and the physical sciences. Problems and answers. 416pp. 5⅜ × 8½. 65251-3 Pa. $10.95

BURNHAM'S CELESTIAL HANDBOOK, Robert Burnham, Jr. Thorough guide to the stars beyond our solar system. Exhaustive treatment. Alphabetical by constellation: Andromeda to Cetus in Vol. 1; Chamaeleon to Orion in Vol. 2; and Pavo to Vulpecula in Vol. 3. Hundreds of illustrations. Index in Vol. 3. 2,000pp. 6⅛ × 9¼. 23567-X, 23568-8, 23673-0 Pa., Three-vol. set $41.85

CHEMICAL MAGIC, Leonard A. Ford. Second Edition, Revised by E. Winston Grundmeier. Over 100 unusual stunts demonstrating cold fire, dust explosions, much more. Text explains scientific principles and stresses safety precautions. 128pp. 5⅜ × 8½. 67628-5 Pa. $5.95

AMATEUR ASTRONOMER'S HANDBOOK, J.B. Sidgwick. Timeless, comprehensive coverage of telescopes, mirrors, lenses, mountings, telescope drives, micrometers, spectroscopes, more. 189 illustrations. 576pp. 5⅜ × 8¼. (Available in U.S. only) 24034-7 Pa. $9.95

CATALOG OF DOVER BOOKS

SPECIAL FUNCTIONS, N.N. Lebedev. Translated by Richard Silverman. Famous Russian work treating more important special functions, with applications to specific problems of physics and engineering. 38 figures. 308pp. 5⅜ × 8½.
60624-4 Pa. $8.95

OBSERVATIONAL ASTRONOMY FOR AMATEURS, J.B. Sidgwick. Mine of useful data for observation of sun, moon, planets, asteroids, aurorae, meteors, comets, variables, binaries, etc. 39 illustrations. 384pp. 5⅜ × 8¼. (Available in U.S. only) 24033-9 Pa. $8.95

INTEGRAL EQUATIONS, F.G. Tricomi. Authoritative, well-written treatment of extremely useful mathematical tool with wide applications. Volterra Equations, Fredholm Equations, much more. Advanced undergraduate to graduate level. Exercises. Bibliography. 238pp. 5⅜ × 8½. 64828-1 Pa. $7.95

POPULAR LECTURES ON MATHEMATICAL LOGIC, Hao Wang. Noted logician's lucid treatment of historical developments, set theory, model theory, recursion theory and constructivism, proof theory, more. 3 appendixes. Bibliography. 1981 edition. ix + 283pp. 5⅜ × 8½. 67632-3 Pa. $8.95

MODERN NONLINEAR EQUATIONS, Thomas L. Saaty. Emphasizes practical solution of problems; covers seven types of equations. ". . . a welcome contribution to the existing literature. . . ."—*Math Reviews*. 490pp. 5⅜ × 8½. 64232-1 Pa. $11.95

FUNDAMENTALS OF ASTRODYNAMICS, Roger Bate et al. Modern approach developed by U.S. Air Force Academy. Designed as a first course. Problems, exercises. Numerous illustrations. 455pp. 5⅜ × 8½. 60061-0 Pa. $9.95

INTRODUCTION TO LINEAR ALGEBRA AND DIFFERENTIAL EQUATIONS, John W. Dettman. Excellent text covers complex numbers, determinants, orthonormal bases, Laplace transforms, much more. Exercises with solutions. Undergraduate level. 416pp. 5⅜ × 8½. 65191-6 Pa. $9.95

INCOMPRESSIBLE AERODYNAMICS, edited by Bryan Thwaites. Covers theoretical and experimental treatment of the uniform flow of air and viscous fluids past two-dimensional aerofoils and three-dimensional wings; many other topics. 654pp. 5⅜ × 8½. 65465-6 Pa. $16.95

INTRODUCTION TO DIFFERENCE EQUATIONS, Samuel Goldberg. Exceptionally clear exposition of important discipline with applications to sociology, psychology, economics. Many illustrative examples; over 250 problems. 260pp. 5⅜ × 8½. 65084-7 Pa. $7.95

LAMINAR BOUNDARY LAYERS, edited by L. Rosenhead. Engineering classic covers steady boundary layers in two- and three-dimensional flow, unsteady boundary layers, stability, observational techniques, much more. 708pp. 5⅜ × 8½.
65646-2 Pa. $18.95

LECTURES ON CLASSICAL DIFFERENTIAL GEOMETRY, Second Edition, Dirk J. Struik. Excellent brief introduction covers curves, theory of surfaces, fundamental equations, geometry on a surface, conformal mapping, other topics. Problems. 240pp. 5⅜ × 8½. 65609-8 Pa. $7.95

ROTARY-WING AERODYNAMICS, W.Z. Stepniewski. Clear, concise text covers aerodynamic phenomena of the rotor and offers guidelines for helicopter performance evaluation. Originally prepared for NASA. 537 figures. 640pp. 6⅛ × 9¼.
64647-5 Pa. $15.95

DIFFERENTIAL GEOMETRY, Heinrich W. Guggenheimer. Local differential geometry as an application of advanced calculus and linear algebra. Curvature, transformation groups, surfaces, more. Exercises. 62 figures. 378pp. 5⅜ × 8½.
63433-7 Pa. $8.95

INTRODUCTION TO SPACE DYNAMICS, William Tyrrell Thomson. Comprehensive, classic introduction to space-flight engineering for advanced undergraduate and graduate students. Includes vector algebra, kinematics, transformation of coordinates. Bibliography. Index. 352pp. 5⅜ × 8½.   65113-4 Pa. $8.95

A SURVEY OF MINIMAL SURFACES, Robert Osserman. Up-to-date, in-depth discussion of the field for advanced students. Corrected and enlarged edition covers new developments. Includes numerous problems. 192pp. 5⅜ × 8½.
64998-9 Pa. $8.95

ANALYTICAL MECHANICS OF GEARS, Earle Buckingham. Indispensable reference for modern gear manufacture covers conjugate gear-tooth action, gear-tooth profiles of various gears, many other topics. 263 figures. 102 tables. 546pp. 5⅜ × 8½.   65712-4 Pa. $14.95

SET THEORY AND LOGIC, Robert R. Stoll. Lucid introduction to unified theory of mathematical concepts. Set theory and logic seen as tools for conceptual understanding of real number system. 496pp. 5⅜ × 8¼.   63829-4 Pa. $10.95

A HISTORY OF MECHANICS, René Dugas. Monumental study of mechanical principles from antiquity to quantum mechanics. Contributions of ancient Greeks, Galileo, Leonardo, Kepler, Lagrange, many others. 671pp. 5⅜ × 8½.
65632-2 Pa. $14.95

FAMOUS PROBLEMS OF GEOMETRY AND HOW TO SOLVE THEM, Benjamin Bold. Squaring the circle, trisecting the angle, duplicating the cube: learn their history, why they are impossible to solve, then solve them yourself. 128pp. 5⅜ × 8½.   24297-8 Pa. $4.95

MECHANICAL VIBRATIONS, J.P. Den Hartog. Classic textbook offers lucid explanations and illustrative models, applying theories of vibrations to a variety of practical industrial engineering problems. Numerous figures. 233 problems, solutions. Appendix. Index. Preface. 436pp. 5⅜ × 8½.   64785-4 Pa. $10.95

CURVATURE AND HOMOLOGY, Samuel I. Goldberg. Thorough treatment of specialized branch of differential geometry. Covers Riemannian manifolds, topology of differentiable manifolds, compact Lie groups, other topics. Exercises. 315pp. 5⅜ × 8½.   64314-X Pa. $8.95

HISTORY OF STRENGTH OF MATERIALS, Stephen P. Timoshenko. Excellent historical survey of the strength of materials with many references to the theories of elasticity and structure. 245 figures. 452pp. 5⅜ × 8½. 61187-6 Pa. $11.95

matrices which define 'f', 'g', and 'c' contain no *identity statements* involving objects of any level higher than 0. In Section 8 Weyl explains why this is of importance. Weyl claims that the reckless use of '=' in the construction of relations can bring impredicativity down on our heads as swiftly as the reckless use of quantification can. Let me try to explain this claim. Weyl regards relations-in-extension as ideal objects which are no less than two stages removed from immediate givenness: they emerge for us only after we have performed acts of both *construction* and *abstraction*. That is, we first mark out a domain of relations-in-intension by defining a construction process for the corresponding sentence matrices. Then we adopt a certain "point of view" toward this domain which renders extensionally equivalent relations-in-intension indiscernable and thus, as far as we can tell from this point of view, identical. We then see before us a domain not of relations-in-intension but of relations-in-extension. As I understand him, Weyl believes that identity claims about relations-in-extension of, say, level 2 are meaningful only when we are able to adopt the abstractive point of view toward a fully *characterized and constituted* domain of individuals and 1st level relations-in-extension. For such identity claims derive their significance from our recognition of the extensional equivalence or non-equivalence of the relevant 2nd level relations-in-intension. And this extensional equivalence or non-equivalence is a matter involving the *totality* of individuals and 1st level relations-in-extension. (Just recall the universal quantifiers at the front of an assertion of extensional equivalence.) If no such totality exists, the identity claims at issue become senseless. Now the question to which all these reflections have been leading is this: Should we posit relations-in-intension of, say, the 3rd level which correspond to sentence matrices containing identity claims about, say, 2nd level relations-in-extension? We should do so only if we are prepared to claim that the domain of individuals and 1st level relations-in-extension can be fully characterized and constituted independently of the 3rd level relations-in-intension we are thinking of positing. For the sentence matrices to which these 3rd level relations correspond are meaningful only if there is such a thing as the totality of individuals and 1st level relations-in-extension. And, thus, if this alleged totality in turn depended on those 3rd level relations, we would be faced with a vicious circle. But it is easy to see that such a circle would in fact obtain. Just let the individual constant 'd' denote a relation-in-extension which corresponds to one of the 3rd level relations-in-intension we are considering. Then the sentence matrix '$x_1^0 \in_1 d$' corresponds to a 1st level relation-in-intension which clearly is dependent upon d for its sense. (Remember: Weyl allows functors and individual constants in the defining clauses of the comprehension axioms in exactly this way.) So the domain of 1st level relations-in-intension and, hence, the domain of 1st level relations-in-extension cannot be fully characterized and constituted independently of d. So if we want to avoid vicious circles, we should *not* posit 3rd level relations-in-intension such as the ones which correspond to d. Accordingly, Weyl adds one final restriction to his comprehension schemes: '=' can appear in ϕ only when it is flanked by variables or constants of level 0. That is, he declines to posit relations which correspond to sentence matrices containing identity claims about any relations-in-extension.

    This concludes our brief tour of Chapter 1. In Chapter 2, Weyl presents his version of analysis in some detail. I think the reader will be able to understand Weyl's presentation perfectly well without any help from me. So the time has come for me to step aside and allow you to explore on your own one of the classics of twentieth-century foundational research.

Stephen Pollard

Northeast Missouri State University

January, 1987

GEOMETRY OF COMPLEX NUMBERS, Hans Schwerdtfeger. Illuminating, widely praised book on analytic geometry of circles, the Moebius transformation, and two-dimensional non-Euclidean geometries. 200pp. 5⅜ × 8¼.

63830-8 Pa. $8.95

MECHANICS, J.P. Den Hartog. A classic introductory text or refresher. Hundreds of applications and design problems illuminate fundamentals of trusses, loaded beams and cables, etc. 334 answered problems. 462pp. 5⅜ × 8½. 60754-2 Pa. $9.95

TOPOLOGY, John G. Hocking and Gail S. Young. Superb one-year course in classical topology. Topological spaces and functions, point-set topology, much more. Examples and problems. Bibliography. Index. 384pp. 5⅜ × 8¼.

65676-4 Pa. $9.95

STRENGTH OF MATERIALS, J.P. Den Hartog. Full, clear treatment of basic material (tension, torsion, bending, etc.) plus advanced material on engineering methods, applications. 350 answered problems. 323pp. 5⅜ × 8½. 60755-0 Pa. $8.95

ELEMENTARY CONCEPTS OF TOPOLOGY, Paul Alexandroff. Elegant, intuitive approach to topology from set-theoretic topology to Betti groups; how concepts of topology are useful in math and physics. 25 figures. 57pp. 5⅜ × 8½.

60747-X Pa. $3.50

ADVANCED STRENGTH OF MATERIALS, J.P. Den Hartog. Superbly written advanced text covers torsion, rotating disks, membrane stresses in shells, much more. Many problems and answers. 388pp. 5⅜ × 8½. 65407-9 Pa. $9.95

COMPUTABILITY AND UNSOLVABILITY, Martin Davis. Classic graduate-level introduction to theory of computability, usually referred to as theory of recurrent functions. New preface and appendix. 288pp. 5⅜ × 8½. 61471-9 Pa. $7.95

GENERAL CHEMISTRY, Linus Pauling. Revised 3rd edition of classic first-year text by Nobel laureate. Atomic and molecular structure, quantum mechanics, statistical mechanics, thermodynamics correlated with descriptive chemistry. Problems. 992pp. 5⅜ × 8½. 65622-5 Pa. $19.95

AN INTRODUCTION TO MATRICES, SETS AND GROUPS FOR SCIENCE STUDENTS, G. Stephenson. Concise, readable text introduces sets, groups, and most importantly, matrices to undergraduate students of physics, chemistry, and engineering. Problems. 164pp. 5⅜ × 8½. 65077-4 Pa. $6.95

THE HISTORICAL BACKGROUND OF CHEMISTRY, Henry M. Leicester. Evolution of ideas, not individual biography. Concentrates on formulation of a coherent set of chemical laws. 260pp. 5⅜ × 8½. 61053-5 Pa. $6.95

THE PHILOSOPHY OF MATHEMATICS: An Introductory Essay, Stephan Körner. Surveys the views of Plato, Aristotle, Leibniz & Kant concerning propositions and theories of applied and pure mathematics. Introduction. Two appendices. Index. 198pp. 5⅜ × 8½. 25048-2 Pa. $7.95

THE DEVELOPMENT OF MODERN CHEMISTRY, Aaron J. Ihde. Authoritative history of chemistry from ancient Greek theory to 20th-century innovation. Covers major chemists and their discoveries. 209 illustrations. 14 tables. Bibliographies. Indices. Appendices. 851pp. 5⅜ × 8½. 64235-6 Pa. $18.95

DE RE METALLICA, Georgius Agricola. The famous Hoover translation of greatest treatise on technological chemistry, engineering, geology, mining of early modern times (1556). All 289 original woodcuts. 638pp. 6¾ × 11.
60006-8 Pa. $18.95

SOME THEORY OF SAMPLING, William Edwards Deming. Analysis of the problems, theory and design of sampling techniques for social scientists, industrial managers and others who find statistics increasingly important in their work. 61 tables. 90 figures. xvii + 602pp. 5⅜ × 8½.
64684-X Pa. $15.95

THE VARIOUS AND INGENIOUS MACHINES OF AGOSTINO RAMELLI: A Classic Sixteenth-Century Illustrated Treatise on Technology, Agostino Ramelli. One of the most widely known and copied works on machinery in the 16th century. 194 detailed plates of water pumps, grain mills, cranes, more. 608pp. 9 × 12.
25497-6 Clothbd. $34.95

LINEAR PROGRAMMING AND ECONOMIC ANALYSIS, Robert Dorfman, Paul A. Samuelson and Robert M. Solow. First comprehensive treatment of linear programming in standard economic analysis. Game theory, modern welfare economics, Leontief input-output, more. 525pp. 5⅜ × 8½.
65491-5 Pa. $14.95

ELEMENTARY DECISION THEORY, Herman Chernoff and Lincoln E. Moses. Clear introduction to statistics and statistical theory covers data processing, probability and random variables, testing hypotheses, much more. Exercises. 364pp. 5⅜ × 8½.
65218-1 Pa. $9.95

THE COMPLEAT STRATEGYST: Being a Primer on the Theory of Games of Strategy, J.D. Williams. Highly entertaining classic describes, with many illustrated examples, how to select best strategies in conflict situations. Prefaces. Appendices. 268pp. 5⅜ × 8½.
25101-2 Pa. $7.95

MATHEMATICAL METHODS OF OPERATIONS RESEARCH, Thomas L. Saaty. Classic graduate-level text covers historical background, classical methods of forming models, optimization, game theory, probability, queueing theory, much more. Exercises. Bibliography. 448pp. 5⅜ × 8¼.
65703-5 Pa. $12.95

CONSTRUCTIONS AND COMBINATORIAL PROBLEMS IN DESIGN OF EXPERIMENTS, Damaraju Raghavarao. In-depth reference work examines orthogonal Latin squares, incomplete block designs, tactical configuration, partial geometry, much more. Abundant explanations, examples. 416pp. 5⅜ × 8¼.
65685-3 Pa. $10.95

THE ABSOLUTE DIFFERENTIAL CALCULUS (CALCULUS OF TENSORS), Tullio Levi-Civita. Great 20th-century mathematician's classic work on material necessary for mathematical grasp of theory of relativity. 452pp. 5⅜ × 8½.
63401-9 Pa. $9.95

VECTOR AND TENSOR ANALYSIS WITH APPLICATIONS, A.I. Borisenko and I.E. Tarapov. Concise introduction. Worked-out problems, solutions, exercises. 257pp. 5⅜ × 8¼.
63833-2 Pa. $7.95

THE FOUR-COLOR PROBLEM: Assaults and Conquest, Thomas L. Saaty and Paul G. Kainen. Engrossing, comprehensive account of the century-old combinatorial topological problem, its history and solution. Bibliographies. Index. 110 figures. 228pp. 5⅜ × 8½. 65092-8 Pa. $6.95

CATALYSIS IN CHEMISTRY AND ENZYMOLOGY, William P. Jencks. Exceptionally clear coverage of mechanisms for catalysis, forces in aqueous solution, carbonyl- and acyl-group reactions, practical kinetics, more. 864pp. 5⅜ × 8½. 65460-5 Pa. $19.95

PROBABILITY: An Introduction, Samuel Goldberg. Excellent basic text covers set theory, probability theory for finite sample spaces, binomial theorem, much more. 360 problems. Bibliographies. 322pp. 5⅜ × 8½. 65252-1 Pa. $8.95

LIGHTNING, Martin A. Uman. Revised, updated edition of classic work on the physics of lightning. Phenomena, terminology, measurement, photography, spectroscopy, thunder, more. Reviews recent research. Bibliography. Indices. 320pp. 5⅜ × 8¼. 64575-4 Pa. $8.95

PROBABILITY THEORY: A Concise Course, Y.A. Rozanov. Highly readable, self-contained introduction covers combination of events, dependent events, Bernoulli trials, etc. Translation by Richard Silverman. 148pp. 5⅜ × 8¼. 63544-9 Pa. $5.95

AN INTRODUCTION TO HAMILTONIAN OPTICS, H. A. Buchdahl. Detailed account of the Hamiltonian treatment of aberration theory in geometrical optics. Many classes of optical systems defined in terms of the symmetries they possess. Problems with detailed solutions. 1970 edition. xv + 360pp. 5⅜ × 8½. 67597-1 Pa. $10.95

STATISTICS MANUAL, Edwin L. Crow, et al. Comprehensive, practical collection of classical and modern methods prepared by U.S. Naval Ordnance Test Station. Stress on use. Basics of statistics assumed. 288pp. 5⅜ × 8½. 60599-X Pa. $6.95

DICTIONARY/OUTLINE OF BASIC STATISTICS, John E. Freund and Frank J. Williams. A clear concise dictionary of over 1,000 statistical terms and an outline of statistical formulas covering probability, nonparametric tests, much more. 208pp. 5⅜ × 8½. 66796-0 Pa. $6.95

STATISTICAL METHOD FROM THE VIEWPOINT OF QUALITY CONTROL, Walter A. Shewhart. Important text explains regulation of variables, uses of statistical control to achieve quality control in industry, agriculture, other areas. 192pp. 5⅜ × 8½. 65232-7 Pa. $7.95

THE INTERPRETATION OF GEOLOGICAL PHASE DIAGRAMS, Ernest G. Ehlers. Clear, concise text emphasizes diagrams of systems under fluid or containing pressure; also coverage of complex binary systems, hydrothermal melting, more. 288pp. 6½ × 9¼. 65389-7 Pa. $10.95

STATISTICAL ADJUSTMENT OF DATA, W. Edwards Deming. Introduction to basic concepts of statistics, curve fitting, least squares solution, conditions without parameter, conditions containing parameters. 26 exercises worked out. 271pp. 5⅜ × 8½. 64685-8 Pa. $8.95

TENSOR CALCULUS, J.L. Synge and A. Schild. Widely used introductory text covers spaces and tensors, basic operations in Riemannian space, non-Riemannian spaces, etc. 324pp. 5⅜ × 8¼. 63612-7 Pa. $8.95

A CONCISE HISTORY OF MATHEMATICS, Dirk J. Struik. The best brief history of mathematics. Stresses origins and covers every major figure from ancient Near East to 19th century. 41 illustrations. 195pp. 5⅜ × 8½. 60255-9 Pa. $7.95

A SHORT ACCOUNT OF THE HISTORY OF MATHEMATICS, W.W. Rouse Ball. One of clearest, most authoritative surveys from the Egyptians and Phoenicians through 19th-century figures such as Grassman, Galois, Riemann. Fourth edition. 522pp. 5⅜ × 8½. 20630-0 Pa. $10.95

HISTORY OF MATHEMATICS, David E. Smith. Nontechnical survey from ancient Greece and Orient to late 19th century; evolution of arithmetic, geometry, trigonometry, calculating devices, algebra, the calculus. 362 illustrations. 1,355pp. 5⅜ × 8½. 20429-4, 20430-8 Pa., Two-vol. set $23.90

THE GEOMETRY OF RENÉ DESCARTES, René Descartes. The great work founded analytical geometry. Original French text, Descartes' own diagrams, together with definitive Smith-Latham translation. 244pp. 5⅜ × 8½.
60068-8 Pa. $6.95

THE ORIGINS OF THE INFINITESIMAL CALCULUS, Margaret E. Baron. Only fully detailed and documented account of crucial discipline: origins; development by Galileo, Kepler, Cavalieri; contributions of Newton, Leibniz, more. 304pp. 5⅜ × 8½. (Available in U.S. and Canada only) 65371-4 Pa. $9.95

THE HISTORY OF THE CALCULUS AND ITS CONCEPTUAL DEVELOPMENT, Carl B. Boyer. Origins in antiquity, medieval contributions, work of Newton, Leibniz, rigorous formulation. Treatment is verbal. 346pp. 5⅜ × 8½.
60509-4 Pa. $8.95

THE THIRTEEN BOOKS OF EUCLID'S ELEMENTS, translated with introduction and commentary by Sir Thomas L. Heath. Definitive edition. Textual and linguistic notes, mathematical analysis. 2,500 years of critical commentary. Not abridged. 1,414pp. 5⅜ × 8½. 60088-2, 60089-0, 60090-4 Pa., Three-vol. set $29.85

GAMES AND DECISIONS: Introduction and Critical Survey, R. Duncan Luce and Howard Raiffa. Superb nontechnical introduction to game theory, primarily applied to social sciences. Utility theory, zero-sum games, n-person games, decision-making, much more. Bibliography. 509pp. 5⅜ × 8½. 65943-7 Pa. $12.95

THE HISTORICAL ROOTS OF ELEMENTARY MATHEMATICS, Lucas N.H. Bunt, Phillip S. Jones, and Jack D. Bedient. Fundamental underpinnings of modern arithmetic, algebra, geometry and number systems derived from ancient civilizations. 320pp. 5⅜ × 8½. 25563-8 Pa. $8.95

CALCULUS REFRESHER FOR TECHNICAL PEOPLE, A. Albert Klaf. Covers important aspects of integral and differential calculus via 756 questions. 566 problems, most answered. 431pp. 5⅜ × 8½. 20370-0 Pa. $8.95

CHALLENGING MATHEMATICAL PROBLEMS WITH ELEMENTARY SOLUTIONS, A.M. Yaglom and I.M. Yaglom. Over 170 challenging problems on probability theory, combinatorial analysis, points and lines, topology, convex polygons, many other topics. Solutions. Total of 445pp. 5⅜ × 8½. Two-vol. set.

Vol. I 65536-9 Pa. $7.95
Vol. II 65537-7 Pa. $6.95

FIFTY CHALLENGING PROBLEMS IN PROBABILITY WITH SOLUTIONS, Frederick Mosteller. Remarkable puzzlers, graded in difficulty, illustrate elementary and advanced aspects of probability. Detailed solutions. 88pp. 5⅜ × 8½.

65355-2 Pa. $4.95

EXPERIMENTS IN TOPOLOGY, Stephen Barr. Classic, lively explanation of one of the byways of mathematics. Klein bottles, Moebius strips, projective planes, map coloring, problem of the Koenigsberg bridges, much more, described with clarity and wit. 43 figures. 210pp. 5⅜ × 8½.

25933-1 Pa. $5.95

RELATIVITY IN ILLUSTRATIONS, Jacob T. Schwartz. Clear nontechnical treatment makes relativity more accessible than ever before. Over 60 drawings illustrate concepts more clearly than text alone. Only high school geometry needed. Bibliography. 128pp. 6⅛ × 9¼.

25965-X Pa. $6.95

AN INTRODUCTION TO ORDINARY DIFFERENTIAL EQUATIONS, Earl A. Coddington. A thorough and systematic first course in elementary differential equations for undergraduates in mathematics and science, with many exercises and problems (with answers). Index. 304pp. 5⅜ × 8½.

65942-9 Pa. $8.95

FOURIER SERIES AND ORTHOGONAL FUNCTIONS, Harry F. Davis. An incisive text combining theory and practical example to introduce Fourier series, orthogonal functions and applications of the Fourier method to boundary-value problems. 570 exercises. Answers and notes. 416pp. 5⅜ × 8½.

65973-9 Pa. $9.95

THE THEORY OF BRANCHING PROCESSES, Theodore E. Harris. First systematic, comprehensive treatment of branching (i.e. multiplicative) processes and their applications. Galton-Watson model, Markov branching processes, electron-photon cascade, many other topics. Rigorous proofs. Bibliography. 240pp. 5⅜ × 8½.

65952-6 Pa. $6.95

AN INTRODUCTION TO ALGEBRAIC STRUCTURES, Joseph Landin. Superb self-contained text covers "abstract algebra": sets and numbers, theory of groups, theory of rings, much more. Numerous well-chosen examples, exercises. 247pp. 5⅜ × 8½.

65940-2 Pa. $7.95
